Synthesis Lectures on Mathematics & Statistics

Series Editor

Steven G. Krantz, Department of Mathematics, Washington University, Saint Louis, MO, USA

This series includes titles in applied mathematics and statistics for cross-disciplinary STEM professionals, educators, researchers, and students. The series focuses on new and traditional techniques to develop mathematical knowledge and skills, an understanding of core mathematical reasoning, and the ability to utilize data in specific applications.

Arturo Portnoy

Calculus to Analysis

An Introductory Transition

 Springer

Arturo Portnoy
Department of Mathematical Sciences
University of Puerto Rico-Mayagüez
Mayagüez, Puerto Rico

ISSN 1938-1743   ISSN 1938-1751 (electronic)
Synthesis Lectures on Mathematics & Statistics
ISBN 978-3-031-69661-9   ISBN 978-3-031-69662-6 (eBook)
https://doi.org/10.1007/978-3-031-69662-6

This Springer imprint is published by the registered company Springer Nature Switzerland AG
The registered company address is: Gewerbestrasse 11, 6330 Cham, Switzerland

If disposing of this product, please recycle the paper.

To my wife Helen, a passionate mathematics and science teacher, much loved and respected by her many students.

Preface

The notes on which this book is based were originally created and designed for mathematics teachers, returning to the university to learn, refresh or relearn more advanced calculus and mathematical analysis, not necessarily in a formal, semester long format. Without claiming to start from the beginning, and without being exhaustive, comprehensive or completely self-contained, this book attempts to serve as a short, concise guide or crash course to review and study the ideas of calculus, of limit and convergence of sequences of functions, and to illustrate the limitations of the Riemann integral and motivate the need for the Lebesgue integral.

A long road is traveled without deviating much from the main route. There are many excellent, comprehensive advanced calculus texts, some of which are listed in the future readings section. These texts contain an abundance of material, and many storylines can be weaved and followed. This is not such a textbook. Here, a single, efficient storyline is proposed: start at the beginning and finish at the end.

Many of the proofs of the presented results are not provided; they are suggested as activities, with occasional hints thrown in to point in the direction of a proof. No lists of exercises are included; the idea is to give a succinct and compact presentation. The tone has been kept informal, even as rigorous arguments are presented and expected. Results, properties, and demonstrations are conveyed in a relaxed prose which hopefully will make reading easier. The included activities frequently use graphing software available online: Desmos Calculator. Some of the activities present a series of notable examples which illustrate the theory; all the suggested activities should be completed. A complementary webpage has been set up,[1] to allow direct play and interactivity with graphs and other demonstrations in the text.

Mathematics teachers in Puerto Rico have very different levels of preparation. An effort was made to address this heterogeneity, which influenced many of the thematic, style, language, depth, rigor and formality choices made.

---

[1] https://calculustoanalysis.weebly.com/.

The hope is that the text will allow readers to efficiently and gently approach advanced calculus and mathematical analysis, and that in a classroom setting, it can be used by instructors to ignite fruitful discussions with their students.

Mayagüez, Puerto Rico   Arturo Portnoy
2024

Contents

1 The Fundamentals 1

2 Sequences, Limits, and Series 15

3 Continuous Functions 25

4 Differentiable Functions 39

5 The Riemann Integral 51

6 Sequences of Functions and Convergence 59

7 Fourier Series I 75

8 Fourier Series II 83

9 Conclusions and Future Directions 91

Appendix A: A Bit on the History of Calculus 95

Appendix B: Heine-Borel Theorem 103

Appendix C: Lebesgue Measure and Integral 107

Further Readings 115

Index 117

List of Figures

Fig. 1.1   Georg Cantor (1845–1918). Attribution: See page for author, Public domain, via Wikimedia Commons, https://commons.wik imedia.org/wiki/File:Georg_Cantor_(Portr%C3%A4t).jpg 6

Fig. 1.2   Seven iterations of Cantor's set construction. Image created with the Desmos graphing calculator, used with permission from Desmos Studio PBC 11

Fig. 2.1   Augustin-Louis Cauchy (1789–1857). Attribution: Public domain, via Wikimedia Commons, https://commons.wikimedia.org/wiki/ File:Augustin-Louis_Cauchy_1901.jpg 17

Fig. 2.2   Bernardus Placidus Johann Nepomuk Bolzano (1781–1848). Attribution: Public Domain, https://commons.wikimedia.org/w/ index.php?curid=73300 17

Fig. 2.3   Karl Theodor Wilhelm Weierstrass (1815–1897). Attribution: Public Domain, https://commons.wikimedia.org/w/index.php? curid=324146 18

Fig. 2.4   Nested intervals. Image created with the Desmos graphing calculator, used with permission from Desmos Studio PBC 19

Fig. 3.1 $\lim_{x \to 1} x^2 = 1$. Image created with the Desmos graphing calculator, used with permission from Desmos Studio PBC 26

Fig. 3.2   Johann Peter Gustav Lejeune Dirichlet (1805–1859). Attribution: By Unknown author—Unknown source, Public Domain, https://commons.wikimedia.org/w/index.php?curid=90476 29

Fig. 3.3   Heaviside step function. Image created with the Desmos graphing calculator, used with permission from Desmos Studio PBC 29

Fig. 3.4 $y = \sin(1/x)$. Image created with the Desmos graphing calculator, used with permission from Desmos Studio PBC 30

Fig. 3.5   Nested intervals. Image created with the Desmos graphing calculator, used with permission from Desmos Studio PBC 32

Fig. 3.6   The fixed point is the intersection of both graphs. Image created
            with the Desmos graphing calculator, used with permission
            from Desmos Studio PBC   34

Fig. 3.7   Finding roots using fixed points. Image created with the Desmos
            graphing calculator, used with permission from Desmos Studio
            PBC   35

Fig. 3.8   Failing to find roots with fixed points. Image created
            with the Desmos graphing calculator, used with permission
            from Desmos Studio PBC   36

Fig. 3.9   Fixing the fixed-point iteration. Image created with the Desmos
            graphing calculator, used with permission from Desmos Studio
            PBC   37

Fig. 4.1   Absolute value function. Image created with the Desmos graphing
            calculator, used with permission from Desmos Studio PBC   40

Fig. 4.2   Derivative of a function. Image created with the Desmos graphing
            calculator, used with permission from Desmos Studio PBC   41

Fig. 4.3   Graph of $f(x)$. Image created with the Desmos graphing
            calculator, used with permission from Desmos Studio PBC   45

Fig. 4.4   Rolle's theorem. Image created with the Desmos graphing
            calculator, used with permission from Desmos Studio PBC   46

Fig. 4.5   Cantor's function. Image created with the Desmos graphing
            calculator, used with permission from Desmos Studio PBC   48

Fig. 5.1   Area under curve. Image created with the Desmos graphing
            calculator, used with permission from Desmos Studio PBC   52

Fig. 5.2   Georg Friedrich Bernhard Riemann (1826–1866). Attribution: By
            http://www.sil.si.edu/digitalcollections/hst/scientific-identity/exp
            lore.htm according to the German Wikipedia, Public Domain,
            https://commons.wikimedia.org/w/index.php?curid=27383   52

Fig. 5.3   The Riemann integral. Image created with the Desmos graphing
            calculator, used with permission from Desmos Studio PBC   54

Fig. 6.1   First six elements in the sequence $f_n(x) = x^n$. Image created
            with the Desmos graphing calculator, used with permission
            from Desmos Studio PBC   60

Fig. 6.2   Brook Taylor (1685–1731). Attribution: Public Domain,
            https://commons.wikimedia.org/w/index.php?curid=524092   63

Fig. 6.3 $f_n(x) = (1 - x^2)^n$. Image created with the Desmos graphing
            calculator, used with permission from Desmos Studio PBC   65

Fig. 6.4 $f_n(x) = n^2 x e^{-nx}$. Image created with the Desmos graphing
            calculator, used with permission from Desmos Studio PBC   67

Fig. 6.5 $f(x) = \sum_{n=0}^{10} \frac{\{10^n x\}}{10^n}$. Image created with the Desmos graphing
            calculator, used with permission from Desmos Studio PBC   70

Fig. 6.6   1st four partial sums of the Weierstrass function. Image created
             with the Desmos graphing calculator, used with permission
             from Desmos Studio PBC 72
Fig. 6.7   11th partial sum of the Weiestrass function. Image created
             with the Desmos graphing calculator, used with permission
             from Desmos Studio PBC 73
Fig. 7.1   Jean-Baptiste Joseph Fourier (1768–1830). Attribution:
             by Julien-Léopold Boilly—This file was derived from:
             Fourier2.jpgRestored by: Bammesk Original *Source* https://www.
             gettyimages.com.au/license/169251384, https://wellcomecollect
             ion.org/works/b4qh352u, Public Domain, https://commons.wik
             imedia.org/w/index.p 76
Fig. 7.2   First partial sums of Fourier series. Image created with the Desmos
             graphing calculator, used with permission from Desmos Studio
             PBC 79
Fig. 7.3   More partial sums of a Fourier series. Image created
             with the Desmos graphing calculator, used with permission
             from Desmos Studio PBC 80
Fig. 7.4   Partial sums of another Fourier series. Image created
             with the Desmos graphing calculator, used with permission
             from Desmos Studio PBC 81
Fig. 8.1   Bessel inequality. The red vector represents $f(x)$, the green vector
             represents $f_\Lambda(x)$, and the orange vector represents their difference
             $f(x) - f_N(x)$. Image created with the Desmos graphing calculator,
             used with permission from Desmos Studio PBC 86
Fig. 8.2   Dirichlet kernel. Image created with the Desmos graphing
             calculator, used with permission from Desmos Studio PBC 89
Fig. 8.3   Fejer kernel. Image created with the Desmos graphing calculator,
             used with permission from Desmos Studio PBC 90
Fig. A.1   Archimedes of Syracuse (287–212 BC). Attribution: By
             Domenico Fetti—http://archimedes2.mpiwg-berlin.mpg.de/archim
             edes_templates/popup.htm, Public Domain, https://commons.wik
             imedia.org/w/index.php?curid=146592 96
Fig. A.2   Method of exhaustion, quadrature of the parabola. Image created
             with the Desmos graphing calculator, used with permission
             from Desmos Studio PBC 97
Fig. A.3   Bonaventura Francesco Cavalieri (1598–1647). Attribution: By
             Unknown author—Trattato della sfera e prattiche per vso di essa,
             Roma, 1682. Public Domain, https://commons.wikimedia.org/w/
             index.php?curid=8546341 98

Fig. A.4   Cavalieri's principle. Image created with the Desmos graphing
            calculator, used with permission from Desmos Studio PBC   99
Fig. A.5   Sir Isaac Newton (1642–1726/27). Attribution: By Godfrey
            Kneller—File: Portrait of Sir Isaac Newton, 1689.jpg from
            https://exhibitions.lib.cam.ac.uk/linesofthought/artifacts, Public
            Domain, https://commons.wikimedia.org/w/index.php?curid=132
            521185   100
Fig. A.6   Gottfried Wilhelm Leibniz (1646–1716). Attribution: By
            Christoph Bernhard Francke—Herzog Anton Ulrich-Museum,
            online, Public Domain, https://commons.wikimedia.org/w/index.
            php?curid=53159699   101
Fig. B.1   Nested intervals. Image created with the Desmos graphing
            calculator, used with permission from Desmos Studio PBC   104
Fig. C.1   Henri Léon Lebesgue (1875–1941). Attribution: Public Domain,
            https://commons.wikimedia.org/w/index.php?curid = 336,482   107
Fig. C.2   Riemann vs. Lebesgue integral. Image created with the Desmos
            graphing calculator, used with permission from Desmos Studio
            PBC   108
Fig. C.3   Iterative construction of Cantor's function. Image created
            with the Desmos graphing calculator, used with permission
            from Desmos Studio PBC   111

The Fundamentals

1

By fundamentals, we mean studying the set of real numbers $\mathbb{R}$. Some familiarity with the natural numbers ($\mathbb{N}$), integers ($\mathbb{Z}$) and rational numbers ($\mathbb{Q}$) is assumed. Let's start with the natural numbers. A very important property of the natural numbers is their well-ordering, which means that any nonempty subset of the natural numbers has a minimal element. On the other hand, we have the principle of mathematical induction, which is used to prove that a proposition $P(n)$ is valid for all $n \in \mathbb{N}$. It states that if

(a) $P(1)$ is true, and
(b) if $P(n)$ is true then $P(n+1)$ is true, for all $n \in \mathbb{N}$,

then $P(n)$ is true for all $n \in \mathbb{N}$.

> **Activity 1.1:**
> Prove that the principle of mathematical induction is equivalent to the well-ordering of the natural numbers.

We know that $\mathbb{N} \subset \mathbb{Z} \subset \mathbb{Q} \subset \mathbb{R}$, and that $\mathbb{Q}$ and $\mathbb{R}$ are closed fields under addition and product.

The following are the field properties of real numbers (and rational numbers) with the usual addition and product:

© The Author(s), under exclusive license to Springer Nature Switzerland AG 2025
A. Portnoy, *Calculus to Analysis*, Synthesis Lectures on Mathematics & Statistics,
https://doi.org/10.1007/978-3-031-69662-6_1

Axioms of addition:

(A1) Closure of addition: if $x, y \in \mathbb{R}$, then $x + y \in \mathbb{R}$.
(A2) Commutativity of addition: if $x, y \in \mathbb{R}$, then $x + y = y + x$.
(A3) Associativity of addition: if $x, y, z \in \mathbb{R}$, then $(x + y) + z = x + (y + z)$.
(A4) Additive neutral: there exists $0 \in \mathbb{R}$ such that if $x \in \mathbb{R}$, then $x + 0 = x$.
(A5) Additive inverse: for each $x \in \mathbb{R}$ there exists $-x \in \mathbb{R}$ such that $x + (-x) = 0$.

Axioms of the product:

(P1) Closure of the product: if $x, y \in \mathbb{R}$, then $xy \in \mathbb{R}$.
(P2) Commutativity of the product: if $x, y \in \mathbb{R}$, then $xy = yx$.
(P3) Associativity of the product: if $x, y, z \in \mathbb{R}$, then $(xy)z = x(yz)$.
(P4) Multiplicative neutral: there exists $1 \in \mathbb{R}$ such that if $x \in \mathbb{R}$, then $1x = x$.
(P5) Multiplicative inverse: for each $x \in \mathbb{R}$ with $x \neq 0$ there exists $x^{-1} \in \mathbb{R}$ such that $xx^{-1} = 1$.

(D) Distributive law: If $x, y, z \in \mathbb{R}$, then $x(y + z) = xy + xz$.

Activity 1.2:
Prove the following propositions for $x, y, z \in \mathbb{R}$:

1. $x + y = x + z \iff y = z$
2. $x + y = 0 \Rightarrow y = -x$ (Uniqueness of the additive inverse)
3. $-(-x) = x$
4. If $x \neq 0$ and $xy = xz$ then $y = z$
5. $xy = 1 \Rightarrow y = x^{-1}$ (Uniqueness of multiplicative inverse)
6. If $x \neq 0$ then $(x^{-1})^{-1} = x$
7. $0x = 0$
8. If $x \neq 0$ and $y \neq 0$ then $xy \neq 0$

Note: All these properties and their consequences are valid in $\mathbb{Q}$ as well.

Why do we need to expand the rational numbers? Since the time of Pythagoras, it has been known that the right triangle of unitary legs has an incommensurable hypotenuse; that is, a hypotenuse that is not a rational number ($\sqrt{2}$). We would like to have a complete number system; that is, a system in which any subset bounded above has a supremum or minimal upper bound within the system. That is, a number greater than or equal to all elements in the subset, but less than or equal to all upper bounds.

Similarly, we expect any subset bounded below to have an infimum or maximal lower bound. That is, a number less than or equal to all elements in the subset, but greater than or equal to all lower bounds.

Let's talk a little bit about the order properties of $\mathbb{R}$. First, we say that $\mathbb{R}$ is an ordered field because:

- Given $x, y \in \mathbb{R}$ then only one of the following propositions is true: $x < y$, $y < x$, $y = x$. (Tricotomy)
- Given $x, y, z \in \mathbb{R}$, if $x < y$ and $y < z$, then $x < z$. (Transitivity)
- Given $x, y, z \in \mathbb{R}$, if $y < z$, then $x + y < x + z$.
- Given $x, y \in \mathbb{R}$, if $x, y > 0$, then $xy > 0$.

Remember that in $\mathbb{R}$ or in $\mathbb{Q}$, we say that $x > y$ or $y < x$ if and only if $x - y > 0$. In addition, we say that $x \geq y$ or $y \leq x$ if and only if $x - y > 0$ or $x - y = 0$. If $x > 0$ we say that x is positive; if $x < 0$ we say that x is negative.

Activity 1.3:
Show that the following propositions are true for $w, x, y, z \in \mathbb{R}$:

- If $a > 0$ and $b > 0$, then $a + b > 0$,
- $x < 0$ if and only if $-x < 0$,
- If $w < x$ and $y < z$, then $w + y < x + z$,
- If $x > 0$ and $y < z$, then $xy < xz$,
- If $x < 0$ and $y < z$, then $xy > xz$,
- If $x \neq 0$ then $x^2 > 0$; in particular $1 > 0$,
- If $0 < x < y$, then $0 < y^{-1} < x^{-1}$.

Activity 1.4:
Consider the set

$$\left\{ \frac{1}{n} \middle| n \in \mathbb{N} \right\} = \left\{ \frac{1}{1}, \frac{1}{2}, \frac{1}{3}, \frac{1}{4}, \cdots \right\}$$

What is the infimum of this set? Is the infimum contained in the set? Find examples of sets of real numbers such that

- the infimum is contained in the set,
- the supremum is not contained in the set,
- the supremum is contained in the set.

Activity 1.5:

(Irrationality of $\sqrt{2}$) Prove that the following bounded subset of the rational numbers has no supremum, or infimal upper bound in the rational numbers:

$$\{x \in \mathbb{Q} : x^2 < 2\}$$

Suggestions:

- Proceed by contradiction: assume that there is a supremum in the rational numbers.
- Show that this supremum must be such that its square is equal to 2. For this, use the following fact: between any two rational numbers there is always another rational number.
- From this a contradiction follows: find it assuming that the numerator and denominator of the supremum are relatively prime.

This suggests that we must extend the rational numbers so that every subset of rational numbers bounded above has a supremum in the extended set. In this case, we would add a number such that when squared it is equal to 2; that is, $\sqrt{2}$.

This is known as the axiom of the supremum, and the extended set is known as the set of real numbers $\mathbb{R}$. The complement of $\mathbb{Q}$ in $\mathbb{R}$ is known as the irrational numbers.

Activity 1.6:

(Archimedean property) Show that if we have two positive real numbers x, y, then there is $n \in \mathbb{N}$ such that $nx > y$.

Suggestions:

- Proceed by contradiction: assume that $nx \leq y$ for all $n \in \mathbb{N}$.
- Therefore, the set $\{nx : n \in \mathbb{N}\}$ is bounded and has a supremum s.

- But $s > s - x$ (why?) and therefore there exists $m \in \mathbb{N}$ such that $mx > s - x$ (why?).
- From the last observation, a contradiction follows.

The Archimedean property has important consequences:

Activity 1.7:
(Density of rational numbers in real numbers) Show that between two different real numbers there is always a rational number, distinct from both.

Suggestions:

- Take the largest of the reals and subtract the smaller. This number is the distance between the two.
- Use the Archimedean property to find a rational less than this distance.
- There will be a natural multiple of this rational that is greater than the smallest of the reals, but less than the larger real number. (This point is the most subtle point of the proof; it is due to the Archimedean property and the well ordering of the naturals.)
- We're done. Why?

Activity 1.8:
(Density of irrational numbers in real numbers) Show that between two different real numbers there is always an irrational, distinct from both.

Suggestions:

- Now, using the Archimedean property find a rational multiple of your favorite irrational that is less than the distance between both numbers.
- There will be a natural multiple of this new number that is greater than the smallest of the reals, but less than the larger real number. (This point is again the most subtle point of the proof; it is due to the Archimedean property and the well ordering of the naturals.)
- We're done. Make sure you understand why.

So, the rational and irrational numbers are very well mixed with each other: there are no open intervals that only contain rational or irrational numbers. All open intervals contain both rational and irrational numbers.

Fig. 1.1 Georg Cantor
(1845–1918). Attribution: See
page for author, Public domain,
via Wikimedia Commons,
https://commons.wikimedia.
org/wiki/File:Georg_Cantor_
(Portr%C3%A4t).jpg

Which are more abundant, the rational or the irrational numbers? This is a delicate question, because both sets are infinite, and deciding which infinite set is larger seems difficult. However, it is a very important question. In fact, one can show that there are as many rational numbers as there are natural numbers (countable infinity), and it is possible to show that the real numbers are uncountable. From this fact one can deduce that irrational numbers are uncountable and therefore much more abundant than rational numbers. These ideas are all due to a Russian-German mathematician named Georg Cantor (1845–1918) (Fig. 1.1).

Cantor's contribution, which we are going to discuss, is a very beautiful and ingenious demonstration of the following proposition: the unit interval is uncountable.

The proof is by contradiction: that is, we will assume that the unit interval is countable, and by logical deduction we will arrive at an obvious contradiction. This will imply that the assumption that originated the contradiction is false, and therefore, its complement true: the unit interval is uncountable.

Let's then assume that the unit interval is countable, and list its elements, listing them in some order below (the order we counted them in), and in decimal representation:

$$0.\, a_1 a_2 a_3 a_4 a_5 \ldots$$
$$0.\, b_1 b_2 b_3 b_4 b_5 \ldots$$
$$0.\, c_1 c_2 c_3 c_4 c_5 \ldots$$
$$0.\, d_1 d_2 d_3 d_4 d_5 \ldots$$
$$\vdots$$

where variables with subscripts represent digits from 0 to 9.

Activity 1.9:

Let's talk a little about the representation of real numbers with respect to a base. What do we mean when we write a number in base 10 (which is the base usually used)? For example, if we write the number 3.1416, this is shorthand notation for:

$$3.1416 = 3 \cdot 10^0 + 1 \cdot 10^{-1} + 4 \cdot 10^{-2} + 1 \cdot 10^{-3} + 6 \cdot 10^{-4}$$

In general, we can represent numbers in base $b \in \mathbb{N}$ with the following notation:
$$...a_2 a_1 a_0.a_{-1}a_{-2}... = ... + a_2 \cdot b^2 + a_1 \cdot b^1 + a_0 \cdot b^0 + a_{-1} \cdot b^{-1} + a_{-2} \cdot b^{-2} + ...$$

where the $a_n \in \{0, 1, 2, ..., b-1\}$. These are called the *digits* of the representation. If a number requires only a finite number of digits in its representation, we say that it is of finite representation.

- Show that if a number is of finite representation, then it is rational.
- Is it true that a rational number always has finite representation?
- Is it true that the representation of a number in a given base is unique?
- Represent 3.1416 in base 2 (binary), base 3, base 7 and base 16.
- Show that a number with infinite but periodic representation (whose digits eventually repeat with a certain period) is rational.

Now, let's build an element of [0,1] not on the list: let its first digit be different than a_1, its second digit different than b_2, its third digit different than c_3, etc. Then, this new item is different from all items listed since it differs by at least one digit from all the items in the list. This is a contradiction. Supposedly the list contained all the elements of [0,1]. Therefore, the unit interval is uncountable, and therefore it is of another category of infinity than that of the set of natural numbers.

In fact, the argument we just made has a small flaw: it assumes that numbers have a unique decimal representation, and that's not true. For example:

$$0.0999999... = 0.1$$

Activity 1.10
Prove that $0.09999\cdots = 0.1$

Suggestions:

- Define the *distance* between two numbers as follows $d(a, b) = |b - a|$.
- Show that this notion of distance satisfies that

  1. $d(a, b) \geq 0$ with equality only if $a = b$,
  2. $d(a, b) = d(b, a)$,
  3. $d(a, b) \leq d(a, c) + d(c, b)$. (Triangle inequality)

- Finally show that $d(0.09999..., 0.1) = 0$, which implies their equality.

The problem with non-unique decimal representation in Cantor's diagonal argument can be solved by including both representations in the original list of numbers, whenever a number admits two representations, which only happens when a number has a finite representation. Another solution would be to force infinite representations for all numbers, including the new number constructed to force the contradiction.

Activity 1.11:
Make sure you understand why non-unique decimal representation causes a logical flaw in the previous argument. How can you make sure that the new number constructed does not coincide with another representation of a number in the list?

Also, let's make sure we understand why rational numbers are countably infinite. In fact, the original argument to prove this is also due to Cantor and goes something like this. Consider the following infinite table of rational numbers:

$\dfrac{1}{1}$ →	↙ $\dfrac{1}{2}$	$\dfrac{1}{3}$ →	↙ $\dfrac{1}{4}$	$\dfrac{1}{5}$ →	...
$\dfrac{2}{1}$ ↓	$\dfrac{2}{2}$ ↗	↙ $\dfrac{2}{3}$	$\dfrac{2}{4}$ ↗	↙ $\dfrac{2}{5}$	...
$\dfrac{3}{1}$ ↗	↙ $\dfrac{3}{2}$	$\dfrac{3}{3}$ ↗	↙ $\dfrac{3}{4}$	$\dfrac{3}{5}$ ↗	...
$\dfrac{4}{1}$ ↓	$\dfrac{4}{2}$ ↗	↙ $\dfrac{4}{3}$	$\dfrac{4}{4}$ ↗	↙ $\dfrac{4}{5}$	...
$\dfrac{5}{1}$ ↗	↙ $\dfrac{5}{2}$	$\dfrac{5}{3}$ ↗	↙ $\dfrac{5}{4}$	$\dfrac{5}{5}$ ↗	...
⋮	⋮	⋮	⋮	⋮	⋱

where the entry in the m th row and n th column is m/n. Notice all positive rational numbers are in the table. Now, lets order or count the elements in the table by starting in the upper left-hand corner (1/1) and following the zig zag pattern of arrows:

$$\frac{1}{1}, \frac{1}{2}, \frac{2}{1}, \frac{3}{1}, \frac{2}{2}, \frac{1}{3}, \frac{1}{4}, \frac{2}{3}, \frac{3}{2}, \frac{4}{1}, \frac{5}{1}, \frac{4}{2}, \frac{3}{3}, \frac{2}{4}, \frac{1}{5}, \cdots$$

Why can we conclude from this observation that the positive rational numbers are countable? Is the fact that the table contains equivalent rational numbers, for example 1/1 and 2/2 a problem? Can we extend this conclusion to all rational numbers (including negative rational numbers)?

Now we will present some fundamental topological notions:

The set $\{x \in \mathbb{R} : a < x < b\}$ is denoted by (a, b) and is called the open interval between a and b.

Let $A \subset \mathbb{R}$ and $x \in A$. We say that x is an interior point of A if there exists an open interval I such that $x \in I \subset A$.

We say that $A \subset \mathbb{R}$ is open if all its elements are interior points of itself.

We say that $A \subset \mathbb{R}$ is closed if $A^c = \mathbb{R} \backslash A$ is open.

Activity 1.12:
Show that the set $\{x \in \mathbb{R} : 0 < x < 1\}$ is open. This set is denoted as (0,1) and is the open interval from 0 to 1.

Activity 1.13:
Show that the set $\{x \in \mathbb{R} : 0 \leq x \leq 1\}$ is closed. This set is denoted as [0,1] and is the closed interval from 0 to 1.

Activity 1.14:
Show that the set $\{x \in \mathbb{R} : 0 < x \leq 1\}$ is neither open nor closed. This set is denoted as (0,1].

Properties of open and closed sets:

- $\mathbb{R}$ is open and closed at the same time. The empty set is also open and closed at the same time.
- The arbitrary union (finite or infinite) of open sets is open.
- The finite intersection of open sets is open.
- The arbitrary intersection (finite or infinite) of closed sets is closed.
- The finite union of closed sets is closed.

Activity 1.15:
Prove the above properties. To prove them you will need to use De Morgan's laws, which are valid for arbitrary unions and intersections of sets. Then find examples of infinite intersections of open intervals that are not open and infinite unions of closed intervals that are not closed.

We say that $A \subset \mathbb{R}$ is compact if it is closed and bounded.

Activity 1.16:
The interval [0,1] is compact. Prove it.

Activity 1.17:
The finite union of closed and bounded intervals is compact. Prove it.

Activity 1.18:
The arbitrary intersection of closed and bounded intervals is compact. Prove it.

If the union of a collection of open sets contains a compact set, then there exists a finite subcollection whose union also contains the compact set. This is the definition of a compact set in many references.

The equivalence between this and the definition we have given in $\mathbb{R}$ is called the Heine-Borel theorem. We refer the reader to Appendix II for a proof of this theorem.

Example: The following example is one of the most famous in the study of mathematical analysis, because it is a relatively simple construction that has surprising properties.

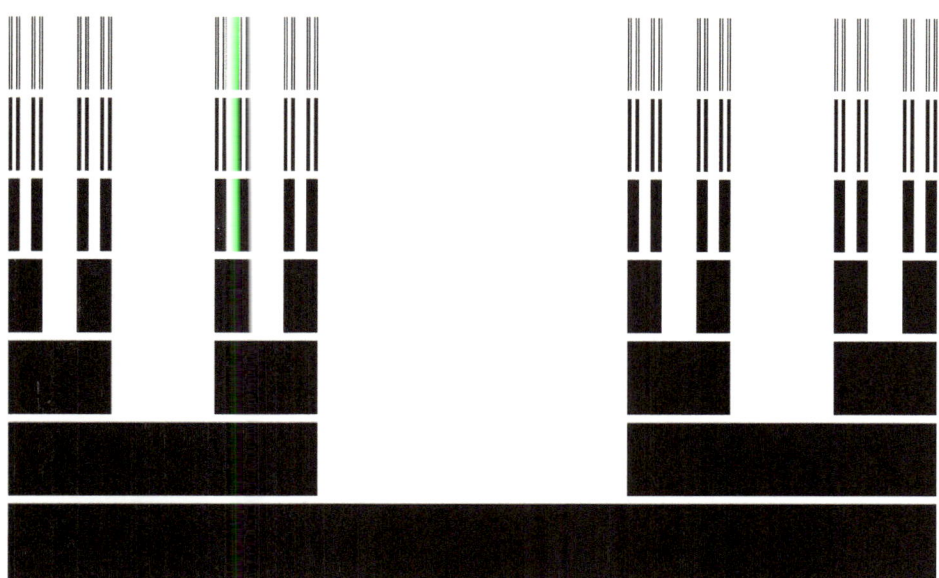

Fig. 1.2 Seven iterations of Cantor's set construction. Image created with the Desmos graphing calculator, used with permission from Desmos Studio PBC

Let's do the construction: We take the interval $C_0 = [0,1]$ and we remove the middle third. What's left is $C_1 = [0,1/3] \cup [2/3,1]$. Now, to the two intervals that remained, we remove the middle third. What's left is $C_2 = [0,1/9] \cup [2/9,1/3] \cup [2/3,7/9] \cup [8/9,1]$. Continuing in this way we construct a nested sequence of closed subsets of the unit interval $C_0 \supset C_1 \supset C_2 \supset \cdots$ (Fig. 1.2):

Let us now consider the intersection of all these sets: $C = \bigcap_{n=0}^{\infty} C_n$. This set is called the Cantor set. The Cantor set is closed and bounded, therefore compact.

Activity 1.19:
Prove that Cantor's set is closed

It is possible to construct the Cantor set in another way. Let's use base 3 to represent numbers in the unit interval,

$$x \in [0,1] \quad x = 0.a_1 a_2 a_3 a_4 \cdots \text{ with } a_n \in \{0, 1, 2\} \text{ for } n \in \mathbb{N}.$$

The Cantor set corresponds to ignoring the numbers that contain 1 as a ternary digit: the elements of the Cantor set are those numbers of the unit interval whose base-3 representation only uses the digits 0 and 2.

Activity 1.20:

Graph all $x = (0.a_1a_2a_3a_4 \ldots)_3$, such that $a_1 \in \{0, 2\}$. Then graph all the $x = (0.a_1a_2a_3a_4 \ldots)_3$, such that $a_1, a_2 \in \{0, 2\}$. Do these sets look familiar?

If you continue with this process, you will obtain a sequence of nested sets, the graphs of which you can observe in Fig. 1.2. If you visit this activity in the book's supplementary website,[1] you can actually see how these graphs were constructed using the base 3 representation of numbers in the interval [0,1] using the Desmos graphing calculator.

Let's define the following function, with domain in the Cantor set:

$$f(0.a_1a_2a_3a_4 \ldots) = (0.(a_1/2)(a_2/2)(a_3/2)(a_4/2) \ldots)_2$$

Note that this function associates to each element of the Cantor set an element of the binary-represented unit interval, and that the range of this function is the entire unit interval. From this we can conclude that the Cantor set, and the unit interval have the same number of elements.

The reader should take a moment to think about the last observation: we have a function that goes from the Cantor set, which is a subset of the unit interval, onto the unit interval: Why does this imply that the two sets have the same "number" of elements?

If $A \subset B \subseteq \mathbb{R}$, and we have an onto function $f(x) : A \to B$, then, because it is surjective, A must have at least one element for each element in B. But because A is a subset of B, A must have the same "number" of elements than B, since it can't have more!

In fact, the two sets are infinite, and in that sense, we cannot count the number of elements they contain. However, we can put in one-to-one correspondence the elements of both sets: for each element of one there is one and exactly one of the other. Can you think of a way to do this (it may be a little tricky, remember that numbers may have more than one representation)? This implies that the sets are the same size or have the same cardinality. The idea of transfinite cardinality, to compare infinite sets, is Cantor's and is one of his most important contributions to set theory and mathematics.

The Cantor set has another very interesting property. Let's measure the length of all the intervals we removed from the unit interval to build it:

$$1/3 + 2/3^2 + 2^2/3^3 + \ldots + 2^n/3^{n+1} + \ldots = 1/3\left(1 + (2/3) + (2/3)^2 + \ldots\right)$$

$$= 1/3\left(\frac{1}{1 - 2/3}\right) = 1.$$

So, we started with an interval of length 1, and removed subintervals with a total length of 1; what remains, the Cantor set, must have zero length or measure.

---

[1] Https://calculustoanalysis.weebly.com/.

How is it possible that two sets with the "same number" of elements, both uncountable, have such different lengths or measures? (The concept of measure will be further discussed later.)

Even though we now know it's true, it contradicts our intuition. But this is precisely why careful analysis is necessary: intuition fails us quickly.

Now we will talk about real number sequences. Real number sequences (which we will simply call sequences from now on) are functions that go from natural numbers to real numbers.

Examples:

$$a_n = 1/n$$

$$b_n = \begin{cases} 1 & \text{if } n \text{ is prime} \\ 0 & \text{otherwise} \end{cases}$$

$$c_n = (-1)^n$$

$$d_n = \sin(n)$$

Notice how a subscript is used to denote the argument of the function. A very important idea is the idea of limit of a sequence: we say that L is the limit of the sequence $\{a_n\}$, if for all $\varepsilon > 0$ there exists an N_ε such that if $n > N_\varepsilon$, then $|L - a_n| < \varepsilon$. In this case, we say that

$$\lim_{n \to \infty} a_n = L$$

© The Author(s), under exclusive license to Springer Nature Switzerland AG 2025
A. Portnoy, *Calculus to Analysis*, Synthesis Lectures on Mathematics & Statistics,
https://doi.org/10.1007/978-3-031-69662-6_2

In the examples above, the sequence $\{a_n\}$ has 0 as its limit, while the sequences $\{b_n\}$, $\{c_n\}$ and $\{d_n\}$ have no limit.

Activity 2.1:
Carefully argue the above observations.

Properties of limits:

- $\lim_{n\to\infty}[\alpha a_n + \beta b_n] = \alpha \lim_{n\to\infty} a_n + \beta \lim_{n\to\infty} b_n$
- $\lim_{n\to\infty}[a_n b_n] = (\lim_{n\to\infty} a_n)(\lim_{n\to\infty} b_n)$
- $\lim_{n\to\infty}[a_n/b_n] = (\lim_{n\to\infty} a_n)/(\lim_{n\to\infty} b_n)$
- If $a_n \leq b_n \leq c_n$ then $\lim_{n\to\infty} a_n \leq \lim_{n\to\infty} b_n \leq \lim_{n\to\infty} c_n$

Note: $\lim_{n\to\infty} a_n$, $\lim_{n\to\infty} b_n$ and $\lim_{n\to\infty} c_n$ must exist, and in the case of the third property, $\lim_{n\to\infty} b_n \neq 0$.

Activity 2.2:
Prove the four properties of limits.

We say that a sequence has Cauchy's property, if for all $\varepsilon > 0$ there exists an N_ε such that if $m, n > N_\varepsilon$, then $|a_m - a_n| < \varepsilon$. It turns out that real sequences with Cauchy's property converge to a limit. In fact, the Cauchy property is a way of characterizing convergent sequences without knowing what they converge to (Fig. 2.1).

An important theorem for sequences is the Bolzano–Weierstrass theorem, which states that any sequence in a compact set has a convergent subsequence in the compact set. In fact, this is one of the fundamental properties that compact sets in $\mathbb{R}$ possess (Figs. 2.2 and 2.3).

Activity 2.3:
Show that the convergence of a sequence is equivalent to the sequence having the Cauchy property.

Suggestions:

- The first implication is simple to prove.
- To prove the second, show that a Cauchy sequence is bounded.
- Next, use the Bolzano–Weierstrass theorem to find a convergent subsequence.

Fig. 2.1 Augustin-Louis
Cauchy (1789–1857).
Attribution: Public domain, via
Wikimedia Commons, https://
commons.wikimedia.org/wiki/
File:Augustin-Louis_Cauchy_
1901.jpg

Fig. 2.2 Bernardus Placicus
Johann Nepomuk Bolzano
(1781–1848). Attribution:
Public Domain, https://com
mons.wikimedia.org/w/index.
php?curid=73300

Fig. 2.3 Karl Theodor
Wilhelm Weierstrass
(1815–1897). Attribution:
Public Domain, https://com
mons.wikimedia.org/w/index.
php?curid=324146

- Finally, show that a Cauchy sequence that has a convergent subsequence must be convergent (to the same limit as the subsequence).

Activity 2.4:
Prove the Bolzano-Weierstrass theorem for sequences in a compact subset of $\mathbb{R}$. The original proof is based on a very famous and ingenious argument. Since we are in a compact set (closed and bounded) in $\mathbb{R}$, We can surround it with a closed and bounded interval. Now, let's take an infinite sequence in that set and therefore in the closed interval. Let's choose an element of the sequence. Now, let's divide the interval into two equal parts. At least in one subinterval there must be an infinite subset of the sequence. Let's choose an element of the sequence that is in that subinterval. On that subinterval we redo the subdivision and choice, and so on (Fig. 2.4):

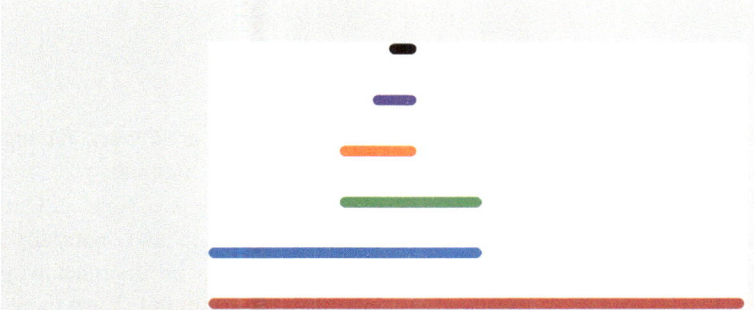

Fig. 2.4 Nested intervals. Image created with the Desmos graphing calculator, used with permission from Desmos Studio PBC

With this process we generate an infinite succession of nested closed subintervals, whose length tends to 0. In addition, the tail of the subsequence that we are selecting falls within the nested subintervals. From these observations it can be argued that the subsequence is a Cauchy subsequence, and therefore converges. Also, since the set is closed, it can be argued that the limit must be inside the set. We leave the details of the proof to the reader.

Note that if a set in $\mathbb{R}$ is not compact, meaning it is not closed or not bounded, then it is easy to find sequences which do not admit a convergent subsequence. Equivalently, if every sequence in a set admits a convergent subsequence, then the set is compact, closed and bounded. This makes the definition of compact and sequentially compact equivalent for sets of real numbers.

Activity 2.5:
A sequentially compact set is one where every sequence in the set has a convergent subsequence whose limit is in the set. Show that compact and sequentially compact are equivalent properties for sets of real numbers.

Now let's talk about series. A series is an infinite sum.

Examples:

$$\sum_{n=1}^{\infty} \frac{1}{n}$$

$$\sum_{n=0}^{\infty} (-1)^n$$

$$\sum_{n=0}^{\infty} \left(\frac{1}{2}\right)^n$$

What does an infinite sum mean? To illustrate the difficulties that arise when talking about adding an infinite number of quantities, consider the following example:

Consider the following infinite sum: $\sum_{n=0}^{\infty}(-1)^n = 1 - 1 + 1 - 1 + 1 - 1 + \dots$. One could associate the terms as follows $(1 - 1) + (1 - 1) + (1 - 1) + \dots$ and "naturally" conclude that the infinite sum is 0. We could also associate the terms in this other way $1 + (-1 + 1) + (-1 + 1) + (-1 + 1) + \dots$ and conclude that the infinite sum is 1. Even more, we could *reorder* the terms of the sum to add n ones and then alternate between positives and negatives (since there are an infinite number of them). In this case, the infinite sum would appear to be equal to n. What is the value of infinite sum then?

One way to give a precise meaning to a series is through its partial sums. If we are considering the series $\sum_{n=1}^{\infty} a_n$, then the partial sums associated with it form the following sequence:

$$s_1 = a_1$$

$$s_2 = a_1 + a_2$$

$$s_3 = a_1 + a_2 + a_3$$

$$\vdots$$

$$s_n = a_1 + a_2 + a_3 + \dots + a_n$$

We will say that the series $\sum_{n=1}^{\infty} a_n$ converges to L, or has L as its limit if $\lim_{n\to\infty} s_n = L$. Clearly, this definition removes the ambiguity of the previous example.

Activity 2.6:
Explain why this definition removes the ambiguity in the previous example.

Example: We will now consider an important infinite sum called the geometric series:

$$s = 1 + r + r^2 + r^3 + r^4 + \dots$$

To study the behavior of this series, let's consider its partial sums:

$$s_n = 1 + r + r^2 + r^3 + r^4 + \cdots + r^n$$

Note that $s_n - rs_n = 1 - r^{n+1}$, and therefore $s_n = \frac{1-r^{n+1}}{1-r}$ (if $r \neq 1$). If $|r| < 1$, then as n grows larger ($n \to \infty$) we have that r^{n+1} tends to 0 ($r^{n+1} \to 0$). Therefore, if $|r| < 1$ we have that s_n tends to $\frac{1}{1-r}$, or $s_n \to \frac{1}{1-r} = s$. This idea of partial sums is very important, because it clearly and unambiguously defines what an infinite sum means: if the partial sums converge to or tend to a value as we add more and more terms, then this is the value or limit of the infinite sum.

Activity 2.7:
Determine whether the following sequence converges, and if it does, what it converges to:

$$\sum_{n=4}^{\infty} 4 \left(\frac{2}{3} \right)^{2n+1}$$

Example: Now consider the harmonic series:

$$1 + \frac{1}{2} + \frac{1}{3} + \frac{1}{4} + \frac{1}{5} + \cdots$$

This example is important, because although the terms we add get smaller and smaller, the partial sums get larger and larger, without bounds. Note that,

$$1 + \frac{1}{2} + \frac{1}{3} + \frac{1}{4} + \frac{1}{5} + \frac{1}{6} + \frac{1}{7} + \frac{1}{8} + \frac{1}{9} + \frac{1}{10} + \frac{1}{11} + \frac{1}{12} + \frac{1}{13} + \frac{1}{14} + \frac{1}{15} + \frac{1}{16} \cdots \geq$$

$$1 + \frac{1}{2} + \frac{1}{4} + \frac{1}{4} + \frac{1}{8} + \frac{1}{8} + \frac{1}{8} + \frac{1}{8} + \frac{1}{16} + \frac{1}{16} + \frac{1}{16} + \frac{1}{16} + \frac{1}{16} + \frac{1}{16} + \frac{1}{16} + \frac{1}{16} \cdots =$$

$$1 + \frac{1}{2} + \frac{1}{2} + \frac{1}{2} + \frac{1}{2} + \cdots$$

That is, $s_{2^n} \geq 1 + n/2$. Therefore, this series does not converge, it diverges.

There are other ways to define convergence of an infinite sum. A famous alternative that will come in handy later is Cesàro summability. We say that an infinite sum converges a la Cesàro or is Cesàro summable if the averages of the first partial sums converge as a sequence. We say then that the series converges to this limit. That is,

$$s = \sum_{n=1}^{\infty} a_n = \lim_{n \to \infty} \sigma_N \text{ where } s_N = \sum_{n=1}^{N} a_n \text{ and } \sigma_N = \frac{1}{N} \sum_{n=1}^{N} s_n$$

It turns out that Cesàro's notion of summability generalizes the classical notion of summability. That is, any convergent series also converges a la Cesàro, and to the same limit. But there are conventionally divergent series that are convergent a la Cesàro.

Example: The series diverges, but let's study it under Cesàro's notion of summability: $\sum_{n=0}^{\infty}(-1)^n$

$$s = \sum_{n=1}^{\infty}(-1)^n = \lim_{n\to\infty}\sigma_N = \frac{1}{2}$$

since

$$s_N = \sum_{n=1}^{N}(-1)^n = \begin{cases} 1 & \text{if } N \text{ is odd} \\ 0 & \text{otherwise} \end{cases}$$

$$\sigma_N = \frac{1}{N}\sum_{n=1}^{N}s_n = \begin{cases} 1/2 & \text{if } N \text{ is even} \\ (N+1)/(2N) & \text{otherwise} \end{cases}$$

So, a conventionally divergent series converges a la Cesàro.

Activity 2.8:
Show that any convergent series converges to the same limit a la Cesàro.

Activity 2.9:
Show that

$$\sum_{n=1}^{\infty}(-1)^n\left(\frac{n^2+1}{n^2}\right)$$

diverges, but converges a la Cesàro. Readers can use the graphing program to convince themselves that the proposition is true.

We say a series $\sum_{n=1}^{\infty}a_n$ converges absolutely if $\sum_{n=1}^{\infty}|a_n|$ converges.
We say a series $\sum_{n=1}^{\infty}a_n$ converges conditionally if it converges, but not absolutely.

Activity 2.10:

Show that the alternating harmonic series $(\sum_{n=1}^{\infty} \frac{(-1)^{n-1}}{n})$ converges conditionally.

Here's an interesting observation: The rearrangement of a series may affect its convergence. More specifically, if the series is conditionally convergent, it may affect the value of its limit.

Activity 2.11:

Show that given any real number S, a rearrangement of the alternating harmonic series exists such that it converges conditionally to S.

Activity 2.12:

(Riemann rearrangement theorem) Show that given any real number S, and any conditionally convergent series of real numbers, a rearrangement of that series exists such that it converges conditionally to S.

Activity 2.13:

Show that any rearrangement of an absolutely convergent series of real numbers converges to the same limit.

Continuous Functions

Now let's consider real functions of a real variable, that is, functions that go from the real numbers to the real numbers. First, we must define what the limit of a function at a point means.

We say that the limit of $f(x) : D \subset \mathbb{R} \to \mathbb{R}$ at $x = a$ is L if and only if for all $\varepsilon > 0$ there exists a $\delta_{\varepsilon,a} > 0$ such that if $x \in D$ and $|x - a| < \delta_{\varepsilon,a}$ then $|f(x) - L| < \varepsilon$. Note that $\delta_{\varepsilon,a}$ depends on ε and on a. In other words, if we get "close enough" to the point where the limit is defined, the image of these points under the function gets arbitrarily close to the limit. "Close enough" depends on how close we want the images to be and where we set the point that defines the limit.

In the definition of limit, we may approach the point of interest at $x = a$ from either side. If we restrict this to approaching only from the left ($x < a$ or $x \to a^-$) or only from the right ($x > a$ or $x \to a^+$), we define the notion of one-sided limits. If a function has both a limit from the right and from the left at $x = a$, and both are equal, this is equivalent to the function having a limit at $x = a$. Can you prove this?

Activity 3.1:
Some interesting observations:

- Note that to prove that $\lim_{x \to a} f(x) \neq L$ it is enough to find a sequence $\{x_n\}$ such that $\lim_{n \to \infty} x_n = a$ and $\lim_{n \to \infty} f(x_n) \neq L$ or $\lim_{n \to \infty} f(x_n)$ does not exist.
- In other words, if $\lim_{x \to a} f(x) = L$ then for any sequence $\{x_n\}$ such that $\lim_{n \to \infty} x_n = a$ we have that $\lim_{n \to \infty} f(x_n) = L$.

(a) Prove the above assertions.
(b) Prove that if a function has both a limit from the right and from the left at $x = a$, and both are equal, this is equivalent to the function having a limit at $x = a$.

© The Author(s), under exclusive license to Springer Nature Switzerland AG 2025
A. Portnoy, *Calculus to Analysis*, Synthesis Lectures on Mathematics & Statistics,
https://doi.org/10.1007/978-3-031-69662-6_3

Activity 3.2:

In the illustration we see the graph of the function $f(x) = x^2$. Prove that

$$\lim_{x \to 1} x^2 = 1$$

See Fig. 3.1.

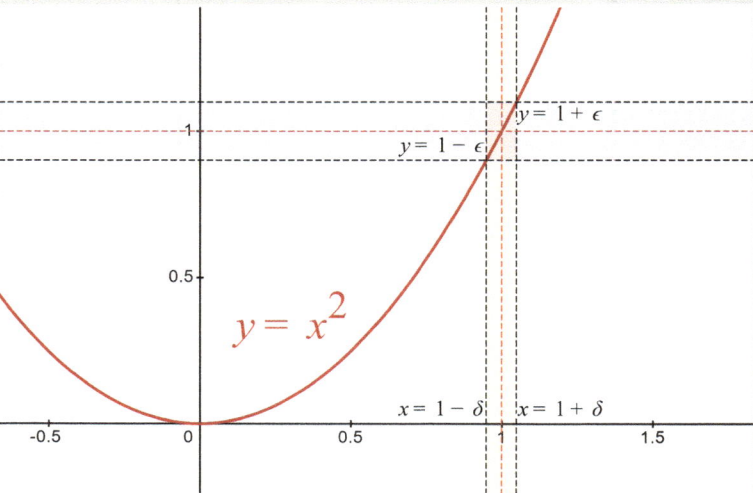

Fig. 3.1 $\lim_{x \to 1} x^2 = 1$. Image created with the Desmos graphing calculator, used with permission from Desmos Studio PBC

Activity 3.3:

Prove that function limits satisfy the same properties as sequence limits:

- $\lim_{x \to a} \left[\alpha f(x) + \beta g(x) \right] = \alpha \lim_{x \to a} f(x) + \beta \lim_{x \to a} g(x)$
- $\lim_{x \to a} (f(x)g(x)) = \left(\lim_{x \to a} f(x) \right) \left(\lim_{x \to a} g(x) \right)$
- $\lim_{x \to a} (f(x)/g(x)) = \left(\lim_{x \to a} f(x) \right) / \left(\lim_{x \to a} g(x) \right)$
- If $f(x) \le g(x) \le h(x)$ close to a, then $\lim_{x \to a} f(x) \le \lim_{x \to a} g(x) \le \lim_{x \to a} h(x)$

Note: $\lim_{x \to a} f(x)$, $\lim_{x \to a} g(x)$ and $\lim_{x \to a} h(x)$ must exist and in the case of the third property, $\lim_{x \to a} g(x) \ne 0$.

Examples:

- $\lim_{x \to 0} xe^{-x} = 0$
- $\lim_{x \to 0} \sin\left(\frac{1}{x}\right)$ does not exist.
- $\lim_{x \to 0} x\sin(1/x) = 0$

Make sure you prove the above examples using the definition of the limit of a function or the properties of function limits. You should also graph each function near $x = 0$ (use the online graphing tool, Desmos) to visualize what is happening.

Activity 3.4:

Prove that

$$\lim_{x \to 0} \frac{\sin(x)}{x} = 1$$

Suggestions:

- Using the unit circle and areas of regions contained within each other, prove that

$$\sin(x) \cos(x) \leq x \leq \tan(x), \text{ for } x > 0$$

- Using the above inequality (and something similar for $x < 0$, and the fourth property of function limits, conclude.

Activity 3.5:

Prove that

$$\lim_{x \to 0} \frac{\cos(x) - 1}{x} = 0$$

Suggestions:

- Multiply and divide the bounding argument by the conjugate of the numerator.
- Use the limit properties to conclude.

Now, we will define continuous functions: We say that a function $f(x) : D \subset \mathbb{R} \to \mathbb{R}$, ranging from a subset of the reals (its domain) to the real numbers, is continuous on $a \in D$ if and only if $\lim_{x \to a} f(x) = f(a)$. In other words, for every $\varepsilon > 0$ there exists a $\delta_{\varepsilon,x} > 0$ such that if $x \in D$ and $|x - a| < \delta_{\varepsilon,x}$ then $|f(x) - f(a)| < \varepsilon$. Note that $\delta_{\varepsilon,x}$ may depend, in general, on both x and ε.

Activity 3.6:
Prove that $f(x) = \sin(x)$ is continuous in $\mathbb{R}$.

Note: To prove that a function is not continuous at $x = a$, it suffices to show that the limit at $x = a$ does not exist or is not equal to $f(a)$, or produce a sequence x_n such that $x_n \to a$, but $|f(x_n) - f(a)| > \varepsilon$ $(f(x_n) \nrightarrow f(a))$.

Example: Consider the following function:

$$f(x) = \begin{cases} 1 & \text{if } x \in \mathbb{Q} \\ 0 & \text{otherwise} \end{cases}$$

This function is known as Dirichlet's function, and it has interesting properties. First, this function is not continuous everywhere (why?). Later we will discuss other interesting properties of this function (Fig. 3.2).

Example: Consider the Heaviside step function:

$$f(x) = \begin{cases} 0 & \text{if } x < 0 \\ 1 & \text{otherwise} \end{cases}$$

The graph of $f(x)$ is shown (Fig. 3.3):
This function is continuous at all points except at $x = 0$, where we have a jump discontinuity, that is, where the limit from the sides at $x = 0$ exist, but are different.

Example: (Thomae) Consider Thomae's function:

$$f(x) = \begin{cases} 1/q & \text{if } x \in \mathbb{Q} \text{ and } x = p/q \text{ is a reduced fraction} \\ 0 & \text{otherwise} \end{cases}$$

This function is continuous on all irrational numbers and discontinuous on all rational numbers.

Fig. 3.2 Johann Peter Gustav Lejeune Dirichlet (1805–1859). Attribution: By Unknown author—Unknown source, Public Domain, https://commons.wikimedia.org/w/index.php?curid=90476

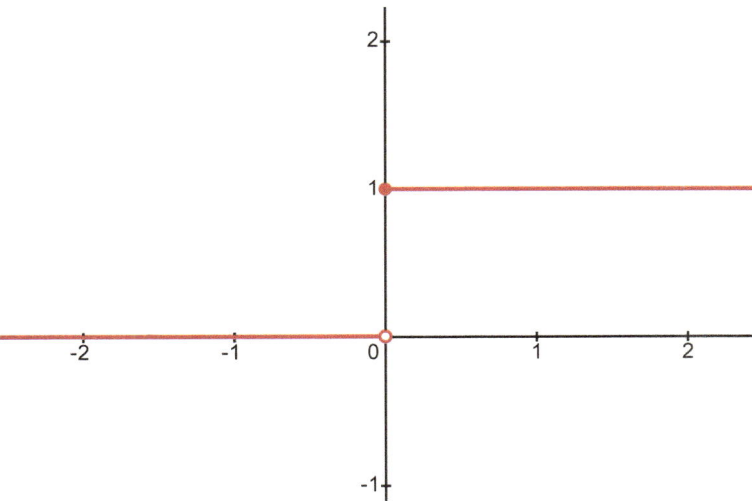

Fig. 3.3 Heaviside step function. Image created with the Desmos graphing calculator, used with permission from Desmos Studio PBC

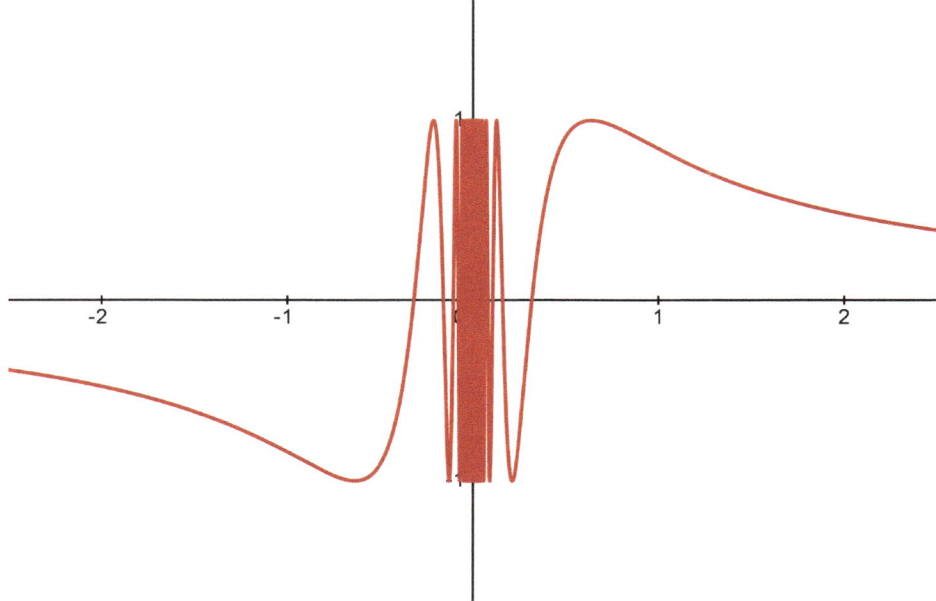

Fig. 3.4 $y = \sin(1/x)$. Image created with the Desmos graphing calculator, used with permission from Desmos Studio PBC

Activity 3.7:

(a) Can you think of a function that is continuous at $x = 0$ and discontinuous everywhere else? Hint: think of Dirichlet's function and modify it.
(b) Show that Thomae's function is continuous on all irrational numbers and discontinuous on all rational numbers.

Activity 3.8:
Prove that the function $f(x) = \sin(1/x)$ for $x \neq 0$ cannot be extended to all reals, so that the extension is continuous in all reals.

We present the graph of $f(x) = \sin(1/x)$ for $x \neq 0$ (Fig. 3.4).

We say that a function $f(x) : D \subset \mathbb{R} \to \mathbb{R}$ is uniformly continuous on D if and only if for every $\varepsilon > 0$ there exists a $\delta_\varepsilon > 0$ such that if $x \in D$ and $|x - a| < \delta_\varepsilon$ then $|f(x) - f(a)| < \varepsilon$. Note that δ_ε depends only on ε and not on a. In other words, if we get "close" enough to any point of continuity, the image of these points under the function arbitrarily approaches the limit. "Close enough" depends only on how much we want the images to come closer, and not on which point of continuity we are working on.

Note: To prove that a continuous function on D is not uniformly continuous on D, it suffices to produce two sequences x_n and a_n in D such that $x_n \to a_n$ or $x_n - a_n \to 0$, but $|f(x_n) - f(a_n)| > \varepsilon$ (or $f(x_n) \nrightarrow f(a_n)$ or $f(x_n) - f(a_n) \nrightarrow 0$).

Activity 3.9:
Consider the following examples,

- Consider the function $f(x) = 2x - 1$ with domain in $\mathbb{R}$. Show that it is uniformly continuous.
- Consider the function $f(x) = 1/x$ with domain in $(0,1]$. Show that it is not uniformly continuous.
- Consider the function $f(x) = x^2$ with domain in $\mathbb{R}$. Show that it is not uniformly continuous. Also show that if we restrict the domain to $[0,3]$, then it is uniformly continuous.
- Consider the function $f(x) = \sin(1/x)$ with domain in $(0,1)$. Show that it is not uniformly continuous.
- Show that $f(x) = \sin(x)$ is uniformly continuous in $\mathbb{R}$.
- We say that a function is Lipschitz continuous in its domain if there exists $M > 0$ such that $|f(x) - f(y)| \le M|x - y|$ for all x, y in its domain. Show that a Lipschitz continuous function is uniformly continuous in its domain.

In the following activities, important properties of continuous functions are explored:

Activity 3.10:
Prove the intermediate value theorem: If a continuous function is defined on an interval $[a, b]$, then it takes on all values between $f(a)$ and $f(b)$ on that interval.

Suggestions:

- The first thing to do is to convince yourself that if the function is not continuous, then the proposition may not be true. Find examples of discontinuous functions that do not satisfy the intermediate value property.
- To prove this property, it is convenient to apply an iterative bisection construction, such as the one we use to prove the Bolzano-Weierstrass theorem for sequences in compact sets. Let C be an intermediate value between $f(a)$ and $f(b)$. We will prove that an element $c \in [a, b]$ exists such that $f(c) = C$.

- Let's take the midpoint between a and b. If $f(x)$ evaluated at that point is C, then we are done. If not, then that value is between $f(a)$ and C or between $f(b)$ and C. In the first case we call the intermediate point a_1 and we call b now b_1. In the second case we call the intermediate point b_1 and we call a now a_1. We repeat this same procedure, but now in the interval $[a_1, b_1]$ to construct the interval $[a_2, b_2]$, and so on. The iterative procedure ends in a finite number of steps (if we find the point) or generates an infinite sequence of closed nested intervals:

$$[a, b] \supset [a_1, b_1] \supset [a_2, b_2] \supset [a_3, b_3] \supset \cdots$$

  See Fig. 3.5.

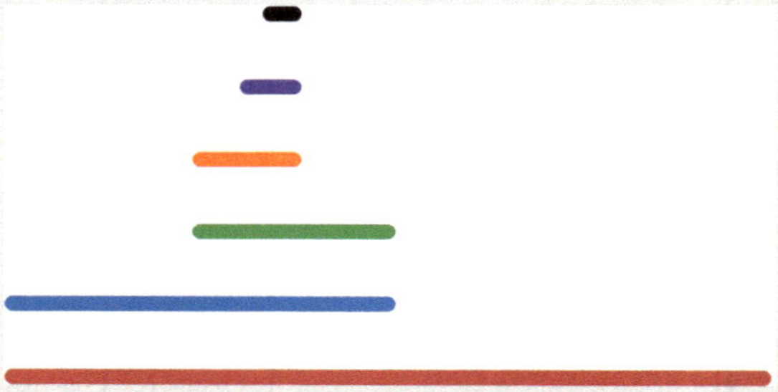

Fig. 3.5 Nested intervals. Image created with the Desmos graphing calculator, used with permission from Desmos Studio PBC

- Their lengths are bisected at each step and therefore tend to zero. This defines two sequences: $\{a_n\}$ and $\{b_n\}$ which tend to the same limit $c \in [a, b]$ (this must be proven, but we leave the details to the reader) and whose images under the function $f(x)$ are some above and some below C. Now, it suffices to prove that $f(c) = C$.
- But $\lim_{n\to\infty} f(a_n) = f(\lim_{n\to\infty} a_n) = f(c) = f(\lim_{n\to\infty} b_n) = \lim_{n\to\infty} f(b_n)$. We have stated that $\{f(a_n)\}$ and $\{f(b_n)\}$ one is above C and one below. The only way this can be true is if both sequences tend to C. We leave the details of this last remark to the reader. This is the second time that the power of the Bolzano-Weierstrass idea becomes evident.
- This algorithm can be used to find solutions of equations or roots, and in numerical analysis it is known as the bisection algorithm. The reader should deduce what is an estimate of the error made by stopping the algorithm at the nth step and using either a_n or b_n as solution estimates to the equation $f(x) = C$.

Activity 3.11:
Consider the equation $x^3 + 2x^2 + x = 3$. Show that it has a solution in the interval [0,1]. Also, find an estimate with double-digit accuracy using the bisection algorithm described above.

Activity 3.12:
Prove that a function $f(x)$ continuous on a compact set reaches its maximum and minimum value in the compact set.

Suggestions:

- The first thing to do is to convince ourselves that the assumptions of the proposition make sense: find examples of discontinuous functions that do not reach their extreme values in a compact set, and examples of continuous functions that do not reach their extreme values on noncompact sets.
- We must convince ourselves that the image of the compact under the continuous function must be bounded. Proceed by contradiction and use the Bolzano–Weierstrass theorem to reach a contradiction.
- Then there is a supremum S and an infimum I for that set image.
- We will show that the function assumes the supremum within the compact set. Let $\{s_n\}$ be a maximizing sequence within the compact set, that is, $\lim_{n \to \infty} f(s_n) = S$ (How do we know this sequence must exist?).
- Since we are in a compact set, the sequence $\{s_n\}$ has a convergent subsequence $\{s_n^*\}$ that converges to s in the compact set. Since $f(x)$ is continuous, $S = \lim_{n \to \infty} f(s_n^*) = f(\lim_{n \to \infty} s_n^*) = f(s)$.
- The proof is analogous for the infimum, and we leave it entirely to the reader.

Activity 3.13:
Prove that a continuous function on a compact set is uniformly continuous.

Suggestions:

- Proceed by contradiction.
- Use the note under the definition of uniform continuity (it says how to prove that a function is not uniformly continuous) and the compactness of the set to arrive at the contradiction.
- Again, Bolzano-Weierstrass to the rescue.

Considering the previous activity, review the examples presented after the definition of uniform continuity.

Activity 3.14:

Prove that a continuous function from a closed interval to itself has at least one fixed point.

Suggestions:

- To begin with, let's clarify what "fixed point" means: x is a fixed point of $f(x)$ if $f(x) = x$. Now, let $f(x) : [a, b] \to [a, b]$. Graphically, the idea is very simple: at a fixed point the graph of $f(x)$ intersects with the line $y = x$ (Fig. 3.6):

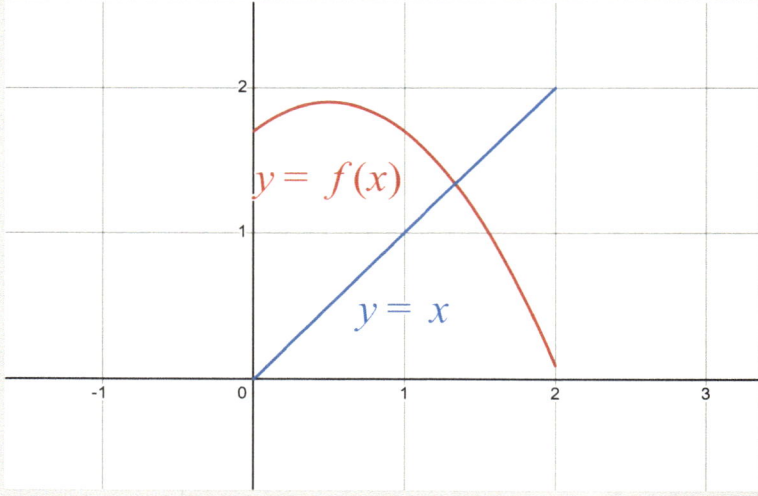

Fig. 3.6 The fixed point is the intersection of both graphs. Image created with the Desmos graphing calculator, used with permission from Desmos Studio PBC

- If $f(x)$ is continuous, its graph must be a continuous curve, making it impossible not to intersect the line $y = x$.
- Consider $g(x) = f(x) - x$. If $g(a) = f(a) - a = 0$, we have our fixed point. Otherwise $g(a) > 0$ (Why?).
- If $g(b) = f(b) - b = 0$, we have our fixed point. Otherwise $g(b) < 0$ (Why?).
- But by the intermediate value theorem for continuous functions there must exist a point $c \in [a, b]$ such that $g(c) = f(c) - c = 0$.

Activity 3.15:

Find examples of functions that go from an open interval to itself and that do not have fixed points, and discontinuous functions from a closed interval to itself that do not have fixed points.

Activity 3.16:

Let's use the idea of a fixed point to solve equations numerically. Suppose we have the following equation: $f(x) = 0$, and that we seek to find solutions to this equation, or roots of $f(x)$. We can reframe the problem in the form of a fixed-point problem in many ways, for example $g(x) = f(x) + x = x$, and we can consider the following functional iteration $x_{n+1} = g(x_n)$, with the hope of generating a sequence that converges to some fixed point. Let's explore this idea with some examples:

- $f(x) = e^{-x} - 1/2 = 0$. Then we would have $g(x) = f(x) + x = e^{-x} - 1/2 + x$, and the functional iteration would be $x_{n+1} = g(x_n)$. Below we present the graph of several iterations corresponding to initial value $x_0 = 2$ (Fig. 3.7):

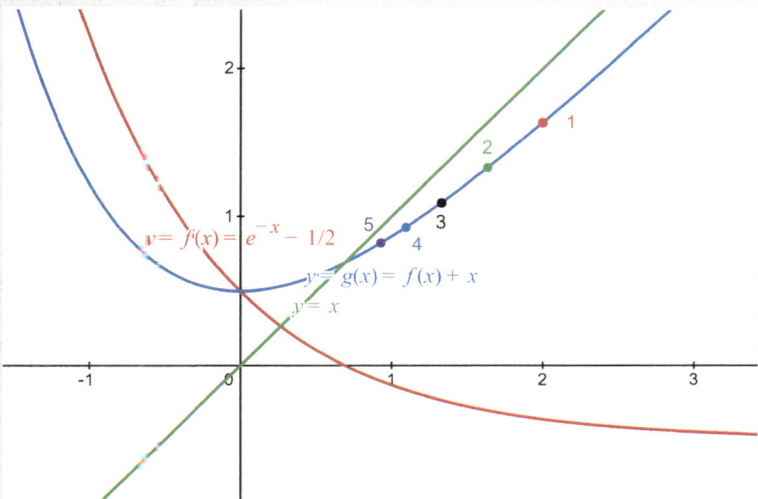

Fig. 3.7 Finding roots using fixed points. Image created with the Desmos graphing calculator, used with permission from Desmos Studio PBC

In this example it seems that any initial condition generates a sequence that converges to the solution or fixed point but, as we shall see, this is not always the case. You can play with this example by changing the initial condition, to verify the claim that any initial condition generates a sequence that converges to the solution, in the book's complementary webpage.[1]

- $f(x) = x^2 - 2 = 0$. Then we would have $g(x) = f(x) + x = x^2 - 2 + x = x$, and the functional iteration would be $x_{n+1} = g(x_n)$. Below we present the graph of an iteration that seems to diverge (Fig. 3.8):

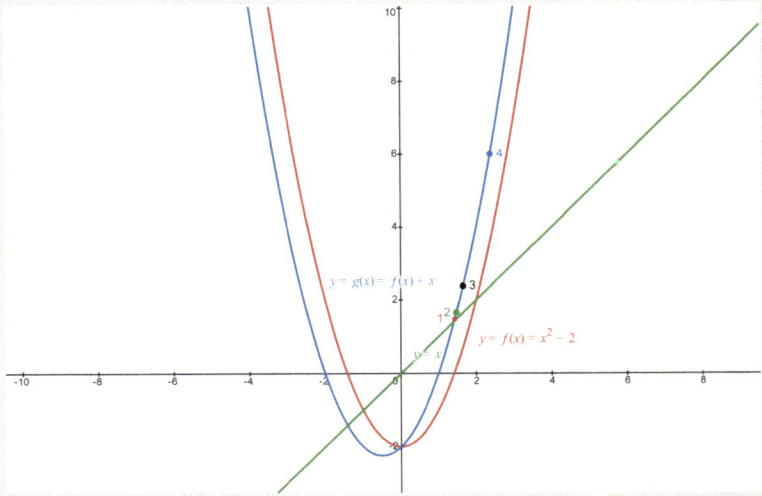

Fig. 3.8 Failing to find roots with fixed points. Image created with the Desmos graphing calculator, used with permission from Desmos Studio PBC

It may be that no iteration seems to generate a convergent sequence. You can play with this example by changing the initial condition, to check if no initial condition generates a sequence that converges to the solution, in the book's complementary webpage.[2]

---

[1] https://calculustoanalysis.weebly.com/.
[2] https://calculustoanalysis.weebly.com/.

- If we slightly modify the previous example: $f(x) = 0.5(x^2 - 2) = 0$. Then we would have $g(x) = f(x) + x = 0.5(x^2 - 2) + x = x$, and the functional iteration would be $x_{n+1} = g(x_n)$. Below we present the graph of several iterations corresponding to an initial value (Fig. 3.9):

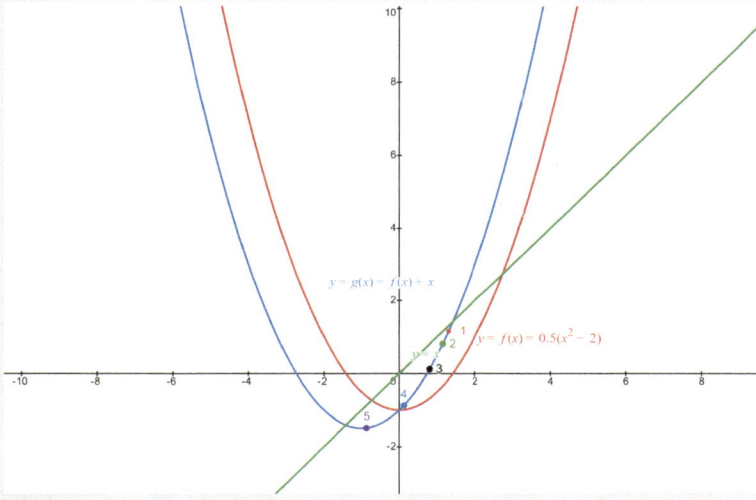

Fig. 3.9 Fixing the fixed-point iteration. Image created with the Desmos graphing calculator, used with permission from Desmos Studio PBC

In this case we see that some initial conditions generate sequences that seem to converge to the solution. $x = -\sqrt{2}$. Do you think some initial conditions will converge to $x = +\sqrt{2}$? Will some initial conditions result in divergent sequences?

Play with these examples in the graphing software until you are convinced of these facts.

We say that a function is a contraction if it is Lipschitz continuous, and its Lipschitz constant is less than 1.

Activity 3.17:
Prove that a contraction going from a closed set $D \subset \mathbb{R}$ to itself, has a single fixed point, and that the sequence defined by the recursion $x_{n+1} = f(x_n)$ converges to this fixed point for any initial value $x_0 \in D$.

- First, prove the uniqueness of the fixed point. Proceed by contradiction.
- Then show that the sequence defined by the recursion is Cauchy, and that it is contained in a closed set, and therefore converges on the set.
- For this point, it should be noted that

$$|x_{n+1} - x_n| = |f(x_n) - f(x_{n-1})| \le L|x_n - x_{n-1}| \le L^n|x_1 - x_0|$$

- This implies that

$$
\begin{aligned}
|x_{n+m} - x_n| &= |x_{n+m} - x_{n+m-1} + x_{n+m-1} - x_{n+m-2} + \ldots + x_{n+1} - x_n| \\
&\le |x_{n+m} - x_{n+m-1}| + |x_{n+m-1} - x_{n+m-2}| + \ldots + |x_{n+1} - x_n| \\
&\le \left(L^{n+m-1} + L^{n+m-2} + \ldots + L^n\right)|x_1 - x_0|
\end{aligned}
$$

- With this and using what we know about the geometric series, the reader should be able to show that the sequence defined by the recursion is Cauchy.
- Finally, use the uniform continuity of Lipschitz functions, and show that the limit of the sequence is the fixed point.
- Consider the examples in activity 3.16 in light of these facts about contractions. Do they shed light on convergence or divergence in those examples?

Differentiable Functions

4

Consider the absolute value function over $[-1,1]$. This is a continuous function, with a "smooth" graph, except at $x = 0$, where we see a "corner" in the graph. We will make these ideas more precise in this chapter where we introduce the concept of the derivative of a function (Fig. 4.1).

We say that a function $f(x) : (a, b) \to \mathbb{R}$ is differentiable at $c \in (a, b)$ if and only if $\lim_{x \to a} \frac{f(x)-f(c)}{x-c}$ exists. In this case, we say that $\frac{d}{dx} f(x)\big|_{x=c} = f'(c) = \lim_{x \to c} \frac{f(x)-f(c)}{x-c} = \lim_{h \to 0} \frac{f(c+h)-f(a)}{h}$ is the derivative of $f(x)$ at $x = c$.

Clearly, the absolute value function is not differentiable at $x = 0$. Make sure you understand why (consider the one-sided limits of $\frac{f(x)-f(c)}{x-c}$ as x tends to c and $c = 0$).

Another way to define a differentiable function is the following. Suppose $f : (a, b) \to \mathbb{R}$ and $c \in (a, b)$. Define $r(h) = f(c+h) - f(c) - Lh$ or write

$$f(c+h) = f(c) + Lh + r(h),$$

that is, the sum of a linear approximation $f(c) + Lh$, plus a residual $r(h)$. If

$$\lim_{h \to 0} \frac{r(h)}{h} = 0$$

then f is differentiable at $x = c$ and the number L represents the derivative of f at $x = c$, $f'(c)$.

You should convince yourself that these are two equivalent ways of defining the derivative of a function at a point. Note that this last definition implies that a differentiable function at $x = c$ can be well approximated (since $r(h) \to 0$ faster than $h \to 0$) by a linear function close to $x = c$.

A. Portnoy, *Calculus to Analysis*, Synthesis Lectures on Mathematics & Statistics,
https://doi.org/10.1007/978-3-031-69662-6_4

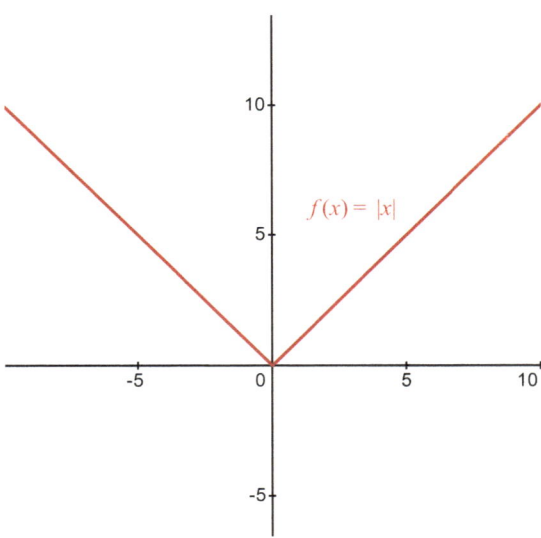

Fig. 4.1 Absolute value function. Image created with the Desmos graphing calculator, used with permission from Desmos Studio PBC

Activity 4.1:

(a) Consider the following propositions:
  • If a function is not continuous at a point, then it is not differentiable at that point.
  • If a function is differentiable at a point, then it is continuous at that point.

Both propositions are equivalent. Can you prove one of them?

(b) Prove the equivalence of both definitions of differentiable function at a point discussed previously (Fig. 4.2).

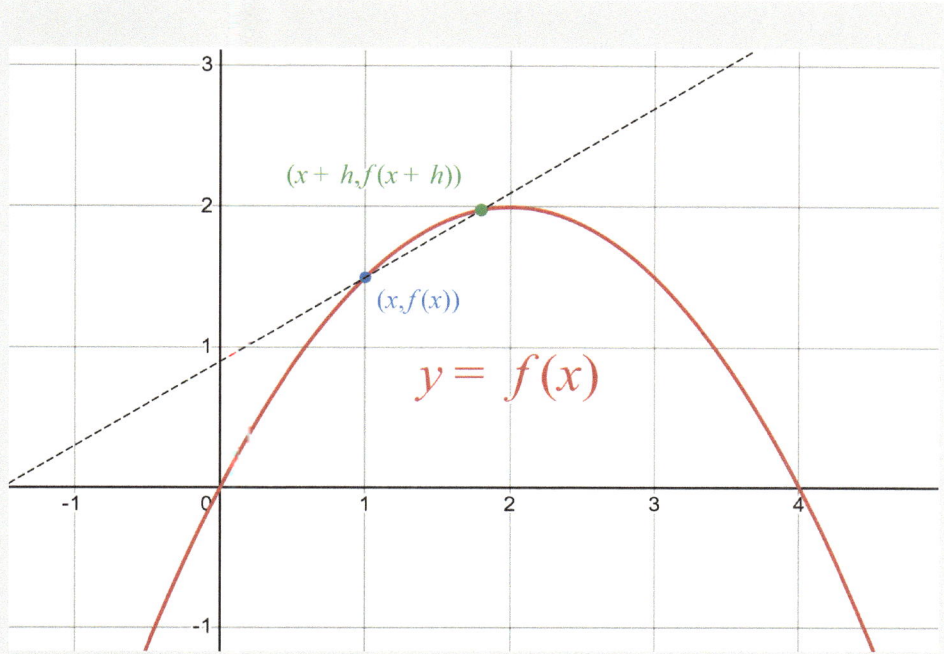

Fig. 4.2 Derivative of a function. Image created with the Desmos graphing calculator, used with permission from Desmos Studio PBC

Activity 4.2:

Prove that $\frac{dx^n}{dx} = nx^{n-1}$, for all $n \in \mathbb{N}$.

Suggestions:

- Start from the definition of derivative as a limit.
- Use Newton's binomial theorem to expand the term $(x+h)^n$.
- Cancel terms in numerator, factor h from each term in the numerator, cancel with h in the denominator, and take the limit as h tends to 0.

Activity 4.3:

Prove that $\frac{d}{dx}a^x \propto a^x$.

Suggestions:

- Start from the definition of derivative as a limit.
- From this it is easy to see that the derivative of a^x is proportional to a^x. The constant of proportionality remains unknown for now.

Activity 4.4:

Prove that $\frac{d}{dx}\sin(x) = \cos(x)$.

Suggestions:

- Start from the definition of derivative as a limit.
- Use the trigonometric identity $\sin(a + b) = \sin(a)\cos(b) + \sin(b)\cos(a)$.
- Use limit properties to conclude.

The derivative of a function has several important interpretations:

- Instantaneous rate of change. By definition, the derivative is the limit of a ratio. In the numerator we have the change of a dependent variable, caused by a change in the independent variable. This change in the independent variable is in the denominator. The classic example is velocity. If we think about $p(t)$ as the position on a horizontal axis of an object as a function of time, then $p'(t) = \lim_{h \to 0} \frac{p(t+h)-p(t)}{h}$ is the change in position divided by the increment of time, taking the limit as this increment tends to 0. This is instantaneous velocity.
- Slope of the tangent line. From the definition of derivative $f'(x) = \lim_{x \to a} \frac{f(x)-p(a)}{x-a}$ we can observe that the quotient $\frac{f(x)-p(a)}{x-a}$ can be interpreted as the slope of a secant line to the graph of $f(x)$. As $x \to a$, the secant line becomes the line tangent to the graph of $f(x)$ at the point $(a, f(a))$.

Activity 4.5:
Consider the function $f(x) = \sin(x)$ and its derivative $f'(x) = \cos(x)$. Plot and interpret them in terms of the instantaneous rate of change and in terms of the slope of the tangent line.

Activity 4.6:
Prove the following properties of the derivative:

- $(\alpha f(x) + \beta g(x))' = \alpha f'(x) + \beta g'(x)$ (linearity of the derivative)
- $(f(x)g(x))' = f'(x)g(x) + g'(x)f(x)$ (product rule)
- $\left(\frac{f(x)}{g(x)}\right)' = \frac{f'(x)g(x) - g'(x)f(x)}{g^2(x)}$ (quotient rule)
- $(f(g(x)))' = f'(g(x))g'(x)$ (chain rule)

Note: $f(x)$ and $g(x)$ must be differentiable. In the case of the third property, $g(x) \neq 0$, and in the case of the fourth property, $f(x)$ must be differentiable over the range of $g(x)$.

Suggestions:

- The property of linearity is proved using properties of the limits.
- To prove the product rule, remember the trick of adding and subtracting some convenient amount from the numerator of the limit argument.
- For the third, assume the existence of the quotient derivative, give it a name, and write the quotient as a product. Then use the product rule.
- For the chain rule, add and subtract, and then multiply and divide by conveniently selected quantities. Finally, use the limit properties.

Activity 4.7:
Prove that $\frac{d}{dx}e^x = e^x$. That is, of all exponentials, the only one whose derivative is exactly equal to itself is the exponential with natural base.

Suggestions:

- Let's start with the traditional definition of the number $e = \lim_{n \to \infty}\left(1 + \frac{1}{n}\right)^n$.
- Note that taking natural logarithms on both sides of the equality, and using the continuity of the logarithm we have $1 = \lim_{n \to \infty} n \ln\left(1 + \frac{1}{n}\right)$.

- Now, let's define $h = \ln\left(1 + \frac{1}{n}\right)$. Then the above limit can be rewritten in terms of h: $1 = \lim_{h \to 0} \frac{h}{e^h - 1}$.
- Using this, conclude.

Activity 4.8:

Use properties of the derivative to demonstrate that $\frac{d}{dx}\ln(x) = \frac{1}{x}$ and that $\frac{d}{dx}\arctan(x) = \frac{1}{1+x^2}$.

Suggestions:

- In both cases use the following: we know the derivative of the inverse function of that which is sought.
- Compose the function whose derivative you search with its inverse, whose derivative you know. This produces the identity function.
- Differentiating both sides of this identity, using the chain rule, we obtain a formula for the derivative of a function in terms of the derivative of its inverse.

Activity 4.9:

Prove that the function $f(x) = \begin{cases} x^2\sin(1/x) & x \neq 0 \\ 0 & x = 0 \end{cases}$ is differentiable in $\mathbb{R}$, but that its derivative is not continuous at $x = 0$. Use a graphing program freely to develop intuition and to visualize what is happening with this function. Below we present the graph of $f(x)$ (Fig. 4.3):

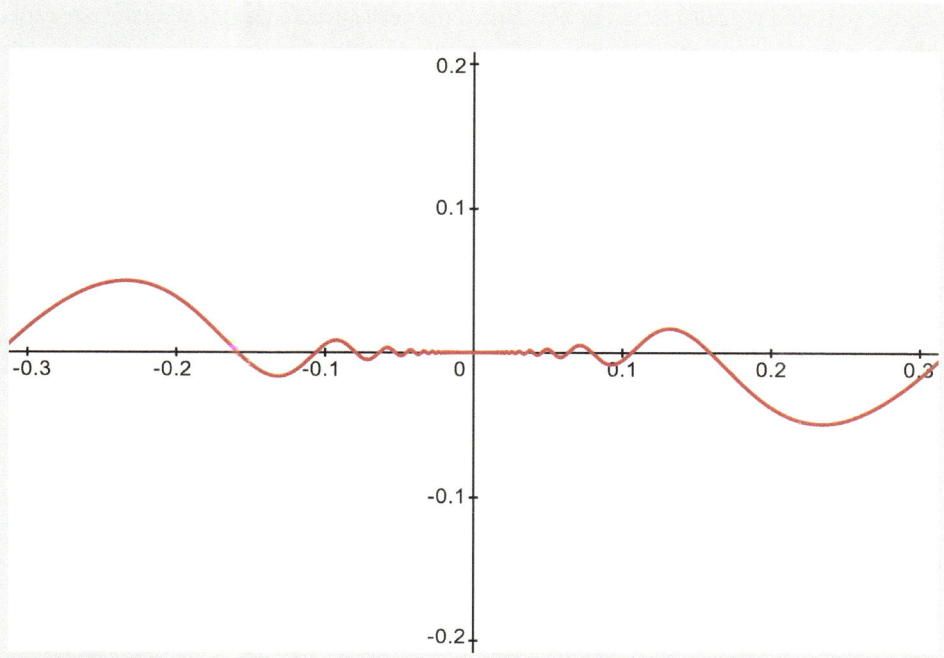

Fig. 4.3 Graph of $f(x)$. Image created with the Desmos graphing calculator, used with permission from Desmos Studio PBC

In the following activities we will prove important properties of differentiable functions:

Activity 4.10:

If $f(x)$ is differentiable at $x = a$ then it is continuous at $x = a$.

Suggestions:

- To prove this property, it is enough to note that what we want to prove is equivalent to $\lim_{x \to a} (f(x) - f(a)) = 0$.
- The proof consists of multiplying and dividing the argument of the limit by $x - a$, use limit properties, and the fact that $f(x)$ is differentiable.

Note: We have seen that differentiable implies continuous, but it turns out that continuous does not imply differentiable. Do you know an example to illustrate this fact?

Note: The Lipschitz constant gives a bound on the derivative and a bound on the derivative gives a Lipschitz constant if the function is differentiable. Can you prove this?

Activity 4.11:

Let $f(x) : [a, b] \to \mathbb{R}$ be a continuous function on $[a, b]$, differentiable on (a, b), and such that $f(a) = f(b)$. Then there exists $c \in (a, b)$ such that $f'(c) = 0$ (Rolle's Theorem) (Fig. 4.4).

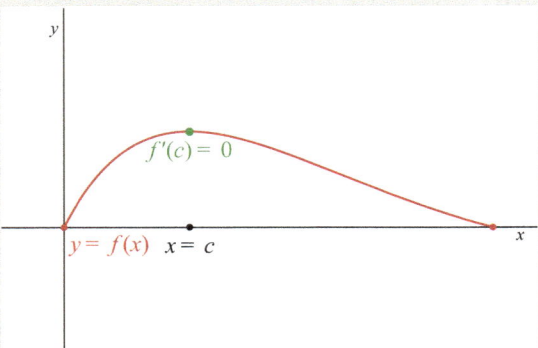

Fig. 4.4 Rolle's theorem. Image created with the Desmos graphing calculator, used with permission from Desmos Studio PBC

Suggestions:

- First, be sure that the assumptions of the theorem make sense. Look for functions that do not satisfy the assumptions and for which the conclusion is not true. Use graphing software to visualize what happens with those examples.
- To prove this result let us first rule out that $f(x)$ is constant, because in that case the result is trivially true.
- If it is not constant, then it assumes its maximum or minimum value at some interior point (Why?); let's call it $c \in (a, b)$.
- At that point $f(x)$ is differentiable (we have assumed this) and its derivative must be 0 (Why?).

Activity 4.12:

Let $f(x) : [a, b] \to \mathbb{R}$ be a continuous function on $[a, b]$, differentiable on (a, b). The there exists $c \in (a, b)$ such that $f'(c)(b-a) = f(b) - f(a)$ (Mean value theorem for derivatives).

Suggestions:

- First, be sure that the assumptions of the theorem make sense. Look for functions that do not satisfy the assumptions and for which the conclusion is not true. Use graphing software to visualize what happens with those examples.
- To prove, consider the function $g(x) = f(x) - f(a) - \frac{f(b)-f(a)}{b-a}(x-a)$. Verify that it satisfies the assumptions of Rolle's theorem and apply it.

Activity 4.13:

Let $f(x) : [0,1] \to [0,1]$ be Cantor's function, otherwise known as the devil's staircase. The Cantor function may be defined as follows:

- First, express x in ternary.
- If the resulting ternary representation contains the digit 1, replace every ternary digit following the 1 by a 0.
- Now replace all 2's with 1's.
- Finally, interpret the result as a binary number which is then $f(x)$.

Remember the construction of the Cantor set, where we used ternary representation as well. We essentially removed all numbers in $[0,1]$ with 1's in their ternary representation. Having refreshed that construction, notice that Cantor's function takes those removed numbers and constructs piecewise constant "steps" with them. For example, ternary points of the form $0.1xxxx\ldots$, which correspond to the first middle third removed in the construction of the Cantor set, are mapped to the binary number 0.1 which is 0.5 in decimal notation (Fig. 4.5).

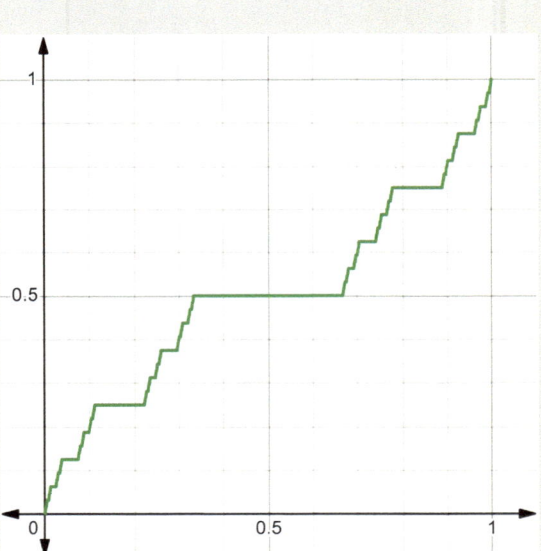

Fig. 4.5 Cantor's function. Image created with the Desmos graphing calculator, used with permission from Desmos Studio PBC

This function is constant "almost everywhere", which implies its derivative is zero "almost everywhere". It is continuous, in fact uniformly continuous, increasing, goes from 0 to 1 and takes every value in between. Amazing!

To end this chapter, we present Newton's method for root finding. Consider the root finding problem $f(x) = 0$. Let's assume that x_0 is near the root x_r and that f is differentiable in an interval containing both x_0 and x_r. Now we approximate $f(x) \approx f(x_0) + f'(x_0)(x - x_0) = 0$ with its linear approximation at x_0, whose graph is a line tangent to the graph of f at x_0, and find the solution of the approximate problem:

$$x_1 = x_0 - \frac{f(x_0)}{f'(x_0)}$$

provided $f'(x_0) \neq 0$. Now we repeat or iterate:

$$x_{n+1} = x_n - \frac{f(x_n)}{f'(x_n)}$$

provided $f'(x_n) \neq 0$. This produces a sequence that may approximate x_r, the true solution of the original equation.

Activity 4.14:

Consider the examples in Activity 3.16.

- Rewrite them as a root finding problem ($g(x) = 0$).
- Make different choices for x_0, run Newton's method and see which sequences converge to each possible solution, which ones diverge, for each example. You should use the Desmos graphing tool for this exploration.

The Riemann Integral

Area under the graph or curve is the classic application to begin the study of the integral (Fig. 5.1).

We will begin our discussion of integrals with some fundamental definitions. A partition of the interval $[a, b]$ is a set:

$$P = \{x_i \in [a, b] \; i = 0, ..., n : a = x_0 \leq x_1 \leq ... \leq x_{n-1} \leq x_n = b\}.$$

We say that another partition Q of the same interval is a refinement of P if $P \subseteq Q$.

Let $f(x) : [a, b] \rightarrow \mathbb{R}$. Then we say that a Riemann sum of the function $f(x)$ corresponding to the partition P is $S(f; P) = \sum_{i=1}^{n} f(x_i^*)\Delta x_i$, where $x_i^* \in [x_{i-1}, x_i]$ and $\Delta x_i = x_i - x_{i-1}$ (Fig. 5.2).

We say that a function $f(x) : [a, b] \rightarrow \mathbb{R}$ is Riemann integrable on $[a, b]$ if a number I exists such that for all $\varepsilon > 0$ there exists a partition of $[a, b]$, let's call it P_ε, such that if Q is any refinement of P_ε, we have that $|I - S(f, Q)| < \varepsilon$. If such a number I exists we say that it is the definite Riemann integral of $f(x)$ over $[a, b]$, and we denote it with this symbol: $\int_a^b f(x)dx$.

An equivalent existence criterion for the Riemann integral is Cauchy's criterion: We say a function $f(x) : [a, b] \rightarrow \mathbb{R}$ is Riemann integrable on $[a, b]$ if for all $\varepsilon > 0$ there exists a partition of $[a, b]$ P_ε such that if Q and R are any two refinements of P_ε, we have that $|S(f, Q) - S(f, R)| < \varepsilon$.

Note: Cauchy's criterion is very useful, because we don't need to know the actual value of the integral to prove integrability.

© The Author(s), under exclusive license to Springer Nature Switzerland AG 2025   51
A. Portnoy, *Calculus to Analysis*, Synthesis Lectures on Mathematics & Statistics,
https://doi.org/10.1007/978-3-031-69662-6_5

Fig. 5.1 Area under curve.
Image created with the Desmos
graphing calculator, used with
permission from Desmos
Studio PBC

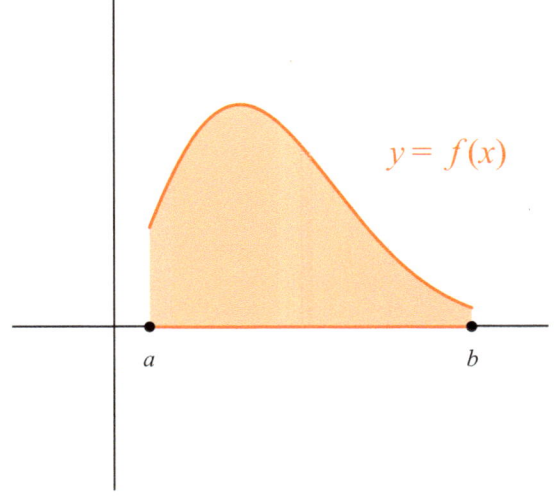

$y = f(x)$

a b

Fig. 5.2 Georg Friedrich
Bernhard Riemann
(1826–1866). Attribution: By
http://www.sil.si.edu/digitalco
llections/hst/scientific-identity/
explore.htm according to the
German Wikipedia, Public
Domain, https://commons.wik
imedia.org/w/index.php?curid=
27383

Activity 5.1:
Show that the Cauchy criterion of existence of the Riemann integral is equivalent to the original definition of integral.

Suggestions:

- Follow the same steps we took in showing that a sequence of real numbers is convergent is equivalent to the sequence having the Cauchy property (Activity 2.3).

Note: It turns out that the continuity of a function ensures the existence of the Riemann integral, regardless of how the x_i^* are chosen.

Activity 5.2:
Prove that the definition of a Riemann definite integral makes sense for continuous functions; that is, the Riemann integral exists (and is the same) no matter how we choose x_i^* provided that the integrand is continuous.

Suggestions:

- Use the fact that a continuous function on a closed and bounded interval is uniformly continuous.
- Use Cauchy's criterion of existence of the integral.

Activity 5.3:
Prove that $\int_0^1 x^2 dx = 1/3$, using the definition of the integral.

Suggestions:

- First, the integrand is continuous, therefore the integral exists.
- Then, use equidistributed partitions: $\Delta x_i = \Delta x = (b - a)/n$.
- To justify why we can use equidistributed partitions, consider the Cauchy criterion and the uniform continuity of the integrand.
- When taking the limit as $n \to \infty$ it will be useful to know that

$$\sum_{i=1}^{n} i^2 = \frac{n(n+1)(2n+1)}{6}$$

Fig. 5.3 The Riemann
integral. Image created with
the Desmos graphing
calculator, used with
permission from Desmos
Studio PBC

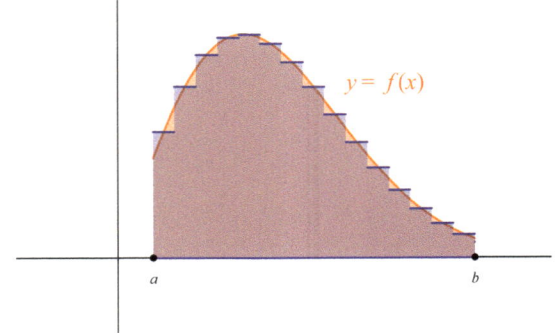

Activity 5.4:

Find an example of a non-Riemann-integrable function. In fact, one of the notable
examples we have discussed is not Riemann integrable.

Suggestion:

- Consider Dirichlet's function. Make sure you understand why it is not Riemann
  integrable.

Activity 5.5:

Consider Thomae's function:

$$f(x) = \begin{cases} 1/q & x \in \mathbb{Q} \text{ and } x = p/q \text{ a reduced fraction} \\ 0 & \text{otherwise} \end{cases}$$

This function is continuous on all irrationals and discontinuous on all rational
numbers. Prove it.

Can you prove that Thomae's function is Riemann integrable on any interval
$[a, b]$?

As was remarked before, the definite integral has an interpretation that follows immediately from its definition, and that is the area under the graph of the integrand. In fact,
a Riemann sum associated to an integral is precisely the approximation of this area by
rectangular strips of base Δx, as the following illustration shows (Fig. 5.3).

If the integrand represents velocity as a function of time, and we integrate over a time
interval $[a, b]$, the interpretation of the integral is of net distance traveled. This physical
interpretation of the integral points to a very important relationship: since velocity is
the time derivative of position, the integral of the derivative of position is net distance

travelled. As we shall soon see, this is the evaluation theorem, part of the fundamental theorem of calculus.

The problem of calculating the area of general regions in the plane is ancient and may have its origins on the land surveying problem, which arose from the advent of agriculture, thousands of years ago. The area problem, related to the integral, historically precedes the slope of the tangent line problem, related to the derivative. Curiously, it is most common to teach the notion of derivative before the integral. We refer the reader to Appendix I, for a bit more on the history of calculus.

Activity 5.6:
Use the Riemann integral definition to justify the interpretation of the integral as net distance travelled.

Suggestions:

- Note that distance = velocity x time.
- If the velocity is positive, one is moving forward. If it is negative, one is moving backward.

We say that a function $F(x)$ is antiderivative of another function $f(x)$ if $F'(x) = f(x)$. It turns out that if $F(x)$ is antiderivative of $f(x)$, then $\int_a^b f(x)dx = F(b) - F(a)$. This is known as the evaluation theorem and is part of the fundamental theorem of calculus. This result gives us an important tool for calculating definite integrals: we look for an antiderivative of the integrand, evaluate at the endpoints of the integration interval and take the difference.

Activity 5.7:
Prove the evaluation theorem.

Suggestions:

- Use the mean-value theorem for derivatives.

Properties of the integral:

- $\int_a^b (\alpha f(x) + \beta g(x))dx = \alpha \int_a^b f(x)dx + \beta \int_a^b g(x)dx$ (linearity).
- $\int_a^b f(x)dx = \int_a^c f(x)dx + \int_c^b f(x)dx$.
- If $f(x) \leq g(x)$ then $\int_a^b f(x)dx = \int_a^b g(x)dx$ (monotonicity).
- Si $f(x) : [a, b] \to \mathbb{R}$ is a continuous function on $[a,b]$, then there exists $c \in [a, b]$ such that $\int_a^b f(x)dx = f(c)(b - a)$ (Mean value theorem for integrals).

Activity 5.8:
Prove the above properties.

Suggestions:

- For the first three, use the definition of integral.
- For the fourth using the evaluation theorem and the mean value theorem for derivatives.

Activity 5.9:
Show that if a function is integrable on an interval, changing the function at a point does not alter either its integrability or the value of the integral on that interval.

A function has many antiderivatives, which differ at most by a constant.

Activity 5.10:
Prove the above observation.

Suggestions:

- First show that given one antiderivative, adding any constant produces another antiderivative. Then show that given any two antiderivatives, they differ by a constant. The mean value theorem for derivatives may be useful.

The indefinite integral of a function $f(x)$ represents any antiderivative of $f(x)$ and is denoted by $\int f(x)dx$.

Now, using the integral we can define new functions:

$$F(x) = \int_a^x f(s)ds$$

It turns out that $F(x)$ is antiderivative of $f(x)$, that is, $F'(x) = f(x)$. This is the second part of the fundamental theorem of calculus.

Activity 5.11:
Prove that $F(x) = \int_a^x f(s)ds$ is antiderivative of $f(x)$. Assume that $f(x)$ is a continuous function.

Suggestions:

- Use the mean-value theorem for integrals.

Activity 5.12:

$\sin(x)$ is antiderivative of $\cos(x)$. Graph both functions with the online graphing tool and study the relationship between their graphs. Make sure you understand the relationship between the area under the graph of $\cos(x)$ and the graph of $\sin(x)$. Also consider that $\cos(x)$ is the derivative of $\sin(x)$. Think of the geometric interpretation of derivative as slope of the tangent line to the graph while looking at the graphs of these two functions.

Other important properties of the integral:

- $\int f'(x)g(x)dx = f(x)g(x) - \int f(x)g'(x)dx$ (Integration by parts)
- $\int f'(g(x))g'(x)dx = f(g(x)) + c$ (Integration by substitution)

Activity 5.13:

Prove the rules of integration by parts and by substitution.

Suggestions:

- For integration by parts, use the rule of differentiation of a product.
- For integration by substitution, use the chain rule.

Activity 5.14:

Use integration by parts or by substitution to find the following integrals:

$\int_0^1 xe^x dx$
$\int_0^1 xe^{x^2} dx$
$\int \ln(x)dx$
$\int_2^4 \frac{\sin(x)}{\sqrt{x}} dx$

We present below the Fundamental Theorem of Calculus. Both parts have already been discussed and proven.

- If $f(x) : [a, b] \to \mathbb{R}$ is a continuous function on $[a,b]$ and we define $F(x) = \int_a^x f(t)dt$ for $x \in [a, b]$, then $F'(x) = f(x)$.
- If $f(x) : [a, b] \to \mathbb{R}$ is a continuous function on $[a,b]$ and $F(x)$ is an antiderivative of $f(x)$, that is $F'(x) = f(x)$, then $\int_a^b f(x)dx = F(x)|_{x=a}^{x=b} = F(b) - F(a)$.

As its name implies, this theorem represents a climactic moment in the development of calculus. On the one hand it relates the Riemann integral with antiderivatives, and on the other it gives us a shortcut to calculate definite integrals without resorting to the clunky definition of integral as the limit of Riemann sums.

Activity 5.15:
Consider Gabriel's horn, the solid of revolution obtained by revolving the graph of $f(x) = 1/x$ for $x \geq 1$ about the x-axis. Note that this defines both a volume and a surface of revolution, both of which are unbounded. We will use this example to review improper integrals, calculating both its surface area and the volume this unbounded region contains.

The volume is given by

$$\int_1^\infty \frac{\pi}{x^2} dx = -\frac{\pi}{x}\Big|_1^\infty = \pi,$$

whereas the surface area is given by

$$\int_1^\infty \frac{2\pi}{x}\sqrt{1 + \frac{1}{x^4}} dx > \int_1^\infty \frac{2\pi}{x} dx = 2\pi \ln(x)\Big|_1^\infty = \infty.$$

If you don't remember how to calculate volumes and surfaces of revolution, this is a good opportunity to review that material. The point though is that (1) the Riemann integral is nice in the sense that improper integrals can be easily defined and calculated and (2) that an unbounded region in space can have a finite volume and a boundary with infinite surface area, which is surprising.

Activity 5.16:
Prove the following version of the Leibniz rule:

$$\frac{d}{dx}\int_{g(x)}^{h(x)} f(t)dt = f(h(x))h'(x) - f(g(x))g'(x)$$

Suggestions:

- Think of the antiderivative of $f(x)$; call it $F(x)$.
- Using the fundamental theorem of calculus, rewrite the integral using $F(x)$, and take the derivative.

Sequences of Functions and Convergence

We will begin our discussion of sequences of functions with several examples.

Example: Consider the following sequence of functions: $f_n(x) = x^n : [0, 1] \rightarrow [0, 1]$. Below we present the graphs of the first six functions in the sequence (Fig. 6.1).

We can see that as $n \rightarrow \infty$, $f_n(x) \rightarrow 0$ if $x \in [0,1)$, but $f_n(1) = 1$ for all n. That is to say that pointwise (thinking about each value of x individually), we have that

$$f_n(x) \rightarrow f(x) = \begin{cases} 0 \text{ if } 0 \leq x < 1 \\ 1 \quad \text{if } x = 1 \end{cases}.$$

Activity 6.1:
Prove the above observation and notice that N_ε depends on x; in fact, the closer x is to 1, the slower $f_n(x)$ tends to 0. You can play around with the Desmos activity which was used to create Fig. 3.4 in the book's complementary webpage,[1] and see what happens as $n \rightarrow \infty$.

We say then that $f_n(x)$ tends or converges pointwise to $f(x)$, or that $f(x)$ is the pointwise limit of $f_n(x)$, and se denote it as $f_n(x) \rightarrow f(x)$. When performing the above activity, one realizes that it is impossible to find a N_ε that works for all x. If it were possible to do so,

---

[1] https://calculustoanalysis.weebly.com/.

© The Author(s), under exclusive license to Springer Nature Switzerland AG 2025
A. Portnoy, *Calculus to Analysis*, Synthesis Lectures on Mathematics & Statistics,
https://doi.org/10.1007/978-3-031-69662-6_6

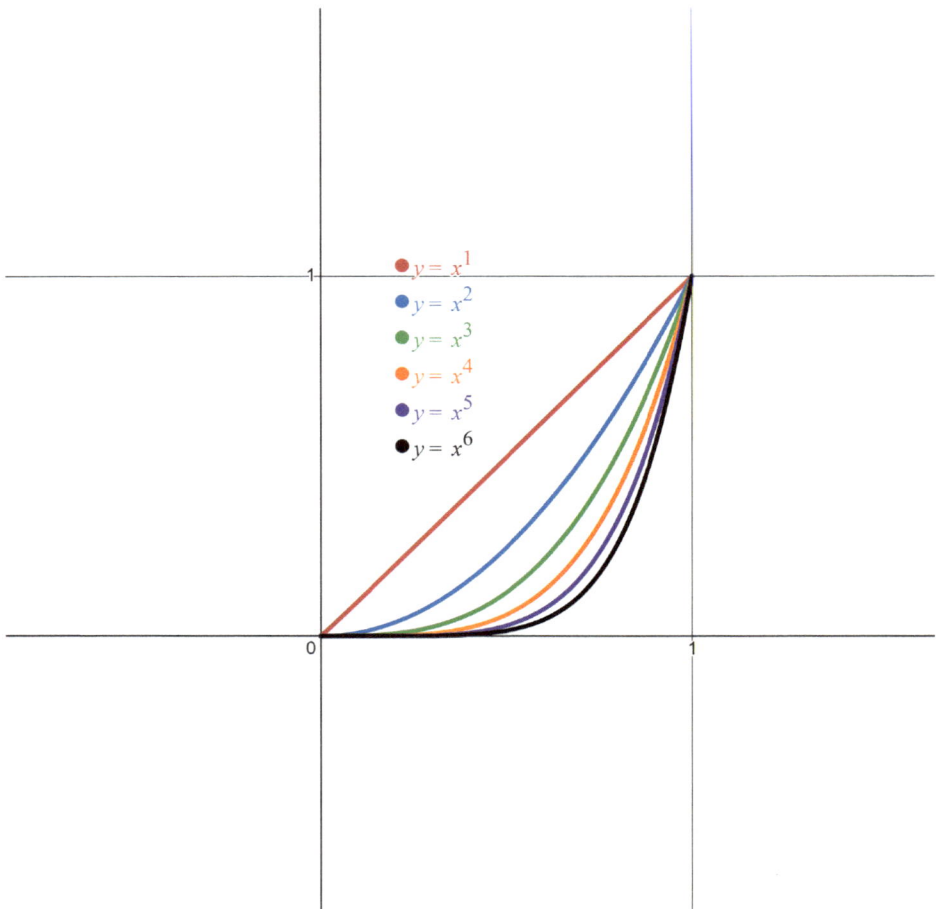

Fig. 6.1 First six elements in the sequence $f_n(x) = x^n$. Image created with the Desmos graphing calculator, used with permission from Desmos Studio PBC

we would say that $f_n(x)$ tends or converges uniformly to $f(x)$ on $[0,1]$, or that $f(x)$ is the uniform limit of $f_n(x)$ on $[0,1]$, and we would denote it as $f_n(x) \xrightarrow{\text{unif}} f(x)$.

Note: Note that in the previous example, a sequence of continuous functions tends pointwise to a function discontinuous at a point.

Activity 6.2:
Prove that $f_n(x) = x^n : [0,0.9] \to \mathbb{R}$ tends uniformly to 0.

Note: As a conclusion to the previous activity, note that by restricting the domain of a sequence of functions, it is possible to strengthen the convergence, from only pointwise to pointwise and uniform. Also, note that in the restricted domain, the limit function is continuous.

Pointwise and uniform convergence properties:

- If $f_n(x) \xrightarrow{unif} f(x)$, then $f_n(x) \to f(x)$, but not vice versa.
- If $f_n(x)$ is a sequence of continuous functions, and $f_n(x) \xrightarrow{unif} f(x)$, then $f(x)$ is continuous in the domain of uniform convergence.

Activity 6.3:
Prove the previous two properties.

Suggestions:

- The first one should be easy.
- For the second, note that:
$$|f(x) - f(a)| = |f(x) - f_n(x) + f_n(x) - f_n(a) + f_n(a) - f(a)|$$
$$\leq |f(x) - f_n(x)| + |f_n(x) - f_n(a)| + |f_n(a) - f(a)|$$
- To bound the first and third terms, use uniform convergence.
- To bound the second term, use the continuity of f_n.

Example: Suppose now that we have a function $f(x) : [a, b] \to \mathbb{R}$ and that $x, y \in (a, b)$. Assuming that $f(x)$ is differentiable and applying the fundamental theorem of calculus we have that

$$f(x) - f(y) = \int_y^x f'(s)ds.$$

Integrating by parts, and assuming that $f(x)$ is $n+1$ times differentiable:

$$f(x) = f(y) + \int_y^x f'(s)ds$$

$$= f(y) + (s-x)f'(s)\Big|_{s=y}^{s=x} - \int_y^x (s-x)f''(s)ds$$

$$= f(y) + (x-y)f'(y) + \frac{(x-y)^2}{2}f''(y) + \cdots$$

$$+ \frac{(x-y)^n}{n!}f^{(n)}(y) + (-1)^n \int_y^x \frac{(s-x)^n}{n!}f^{(n+1)}(s)ds$$

To the sum of the first $n+1$ are called the Taylor polynomial of $f(x)$ of order n centered at y, and we denote it as $T_n^y(x)$. The last term is called the error associated with $T_n^y(x)$, and we denote it as $E_n^y(x)$. If $y = 0$, that is, the Taylor polynomial is centered at $y = 0$, then we call it the Maclaurin polynomial. If this process is continued indefinitely, the result is the Taylor series associated with $f(x)$, and if the series is centered at $y = 0$, it is called the Maclaurin series. Both are power series, since their terms are powers of the variable x.

Observe that

$$\left| E_n^y(x) \right| = \left| (-1)^n \int_y^x \frac{(s-x)^n}{n!}f^{(n+1)}(s)ds \right| \le M \frac{|x-y|^{n+1}}{(n+1)!}$$

if $\left| f^{(n+1)}(s) \right| \le M$ for all s in an interval containing x and y. This is known as Taylor's inequality. Can you produce a detailed proof of the inequality?

Representing or approximating functions with power series is very useful. Power series are essentially long polynomials, which are easy to evaluate (you only need to addition and multiplication to do it), differentiate, and integrate. There are many situations in which it is convenient to replace a "difficult" function with a polynomial that approximates it well (Fig. 6.2).

Activity 6.4:
Prove that if $f(x) = e^x$ on $[-5,5]$, then $T_n^0(x) \to f(x)$, and further that $T_n^0(x) \overset{unif}{\to} f(x)$. Use the online graphing tool to visualize the graphs of $f(x)$ and $T_n^0(x)$ for various values of n.

Note that $T_n^0(1) \to f(1) = e$ provides an approximation (quickly converging, by the way) of the transcendental irrational e.

Fig. 6.2 Brook Taylor
(1685–1731). Attribution:
Public Domain, https://com
mons.wikimedia.org/w/index.
php?curid=524092

Suggestions:

- Note that $f(x) = e^x$ is infinitely differentiable and all its derivatives are equal to $f(x)$.
- Then, the error term can be bounded in that interval by something that tends to zero as $n \to \infty$. Use Taylor's inequality.

Is the convergence uniform on all $\mathbb{R}$?

Activity 6.5:
Repeat the previous activity with the following functions

- $f(x) = \sin(x)$ on $[-2\pi, 2\pi]$
- $f(x) = \cos(x)$ on $[-2\pi, 2\pi]$
- $f(x) = 1 - x + x^2 - x^3$ on $[-3,4]$

Suggestion: Follow a line of reasoning similar to the previous activity.

Activity 6.6:

Figure out the Taylor expansion of $f(x) = \arctan(x)$ centered on $x = 0$ and use it to find approximations of π.

Suggestions:

- Consider the power series expansion of $g(x) = \frac{1}{1+x^2}$.
- Note that $f'(x) = g(x)$. This should yield the sought after Taylor expansion.
- Figure out where this power series converges and think of particular values of x that may be of interest to estimate π.
- You may want to consider the trigonometric identity $\tan(\alpha - \beta) = \frac{\tan(\alpha)-\tan(\beta)}{1+\tan(\alpha)\tan(\beta)}$.

Note: These activities suggest that polynomials closely approximate functions that are very "smooth", or that have many derivatives. Polynomials happen to be good approximations to any continuous function. This was proven by Weierstrass in a famous theorem that bears his name.

(Weierstrass theorem) If $f(x)$ is continuous on an interval $[a, b]$, then there is a sequence of polynomials $P_n(x)$ that converges to $f(x)$ uniformly on $[a, b]$.

Let's discuss the proof of the Weierstrass theorem. First, let's note that without losing generality, we can assume that $[a, b] = [0,1]$ and that $f(0) = f(1) = 0$ (Why?). In addition, we will agree that $f(x)$ is 0 outside of $[0,1]$. In this way we ensure that $f(x)$ is uniformly continuous on $\mathbb{R}$.

Now let's consider the following polynomials: $Q_n(x) = c_n(1 - x^2)^n$, choosing the normalizing constants c_n in such a way that $\int_{-1}^{1} Q_n(x)dx = 1$. Let's plot the first elements of this sequence of functions without the normalizing constants (Fig. 6.3).

What do these polynomials do? They all have area 1 under their graph (with the normalizing constants), but as $n \to \infty$, they concentrate that area around the origin.

Now let's define the following sequence of polynomials:

$$P_n(x) = \int_{-1}^{1} f(x + t)Q_n(t)dt = \int_{-x}^{1-x} f(x + t)Q_n(t)dt = \int_{0}^{1} f(t)Q_n(t - x)dt$$

First you must convince yourself that the three integrals are equivalent. The last one clearly defines a polynomial in x, so we are justified in saying that the sequence is of polynomials. Now let's look at the first integral: if the $Q_n(t)$ concentrate around the origin and integrate to 1, that integral should be very similar to $f(x)$ as $n \to \infty$. The remainder of the proof is based on this last observation and consists of bounding the difference $|P_n(x) - f(x)|$ uniformly:

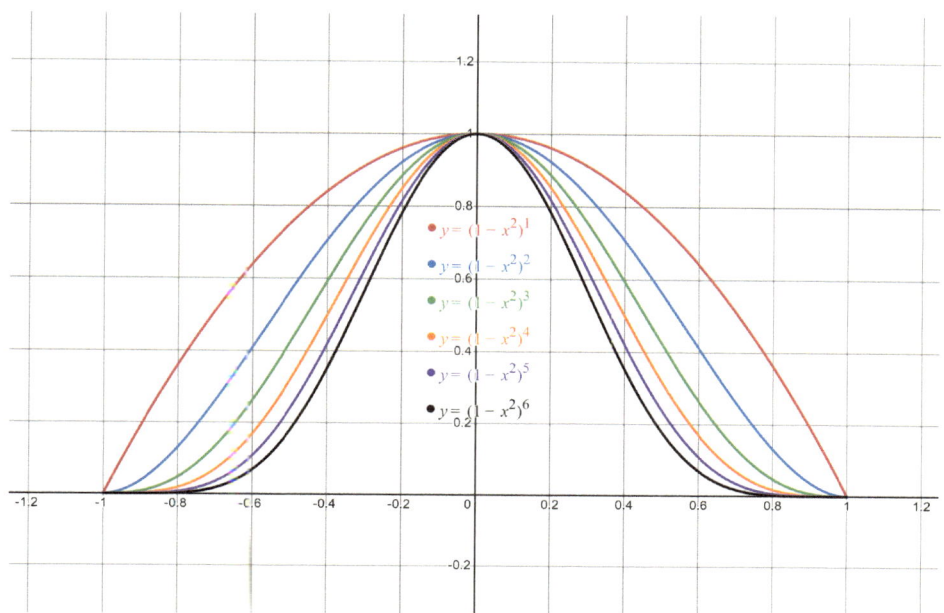

Fig. 6.3 $f_n(x) = (1 - x^2)^n$. Image created with the Desmos graphing calculator, used with permission from Desmos Studio PBC

$$|P_n(x) - f(x)| = \left| \int_{-1}^{1} (f(x+t) - f(x)) Q_n(t) dt \right|$$

$$\leq \int_{-1}^{1} |f(x+t) - f(x)| Q_n(t) dt$$

$$= \int_{-1}^{-\delta} |f(x+t) - f(x)| Q_n(t) dt$$

$$+ \int_{-\delta}^{-\delta} |f(x+t) - f(x)| Q_n(t) dt$$

$$+ \int_{\delta}^{1} |f(x+t) - f(x)| Q_n(t) dt$$

We leave as an exercise for the reader the details of how to bound each of these integrals. The uniform continuity of $f(x)$ is useful in the proof.

Activity 6.7:
Complete the proof of the Weierstrass theorem.

Suggestions:

- Prove that $c_n < \sqrt{n}$.
- With this, bound the first and third terms of inequality with something that tends to zero as $n \to \infty$.
- The second term can be bound using the uniform continuity of $f(x)$.

Following are some interesting examples of function sequences, their limits, and their properties:

Example: Let's define the following sequence of functions over $0 \leq x < \infty$:

$$f_n(x) = n^2 x e^{-nx}$$

The graphs of the first six elements of the sequence are shown below (Fig. 6.4).

It's easy to verify that $\int_0^\infty f_n(x)dx = \int_0^\infty n^2 x e^{-nx}dx = 1$ for all $n \in \mathbb{N}$. We suggest the reader verifies this. Notice that as $n \to \infty$, the graphs lengthen and narrow, that is, they concentrate the area under the curve closer and closer to the origin. On the other hand, as $n \to \infty$, $f_n(x) \to 0$ for any $x > 0$. We can then conclude that pointwise, $f_n(x) \to 0$ for all $x \geq 0$. However, $\lim_{n\to\infty} \int_0^\infty f_n(x)dx = 1$. Therefore $1 = \lim_{n\to\infty} \int_0^\infty f_n(x)dx \neq \int_0^\infty \lim_{n\to\infty} f_n(x)dx = \int_0^\infty 0 dx = 0$. That is, that we can't, in general, swap or exchange a pointwise limit with the Riemann integral, because we can get different results.

Activity 6.8:
Construct another example of a sequence of functions defined over a bounded interval ([0,1] for example), where the exchange of pointwise limit and integral also gives different results. Plot your sequence using the Desmos graphing tool.

Example: Now let's think of the Dirichlet function, but as the limit of a sequence of functions. We know that rational numbers in the interval [0,1] are countable. Let's list them as follows: $r_1, r_2, r_3, \dots$. Now let's define the following sequence of functions:

$$f_n(x) = \begin{cases} 1 & \text{if } x \in \{r_1, r_2, r_3, \dots r_n\} \\ 0 & \text{otherwise} \end{cases}$$

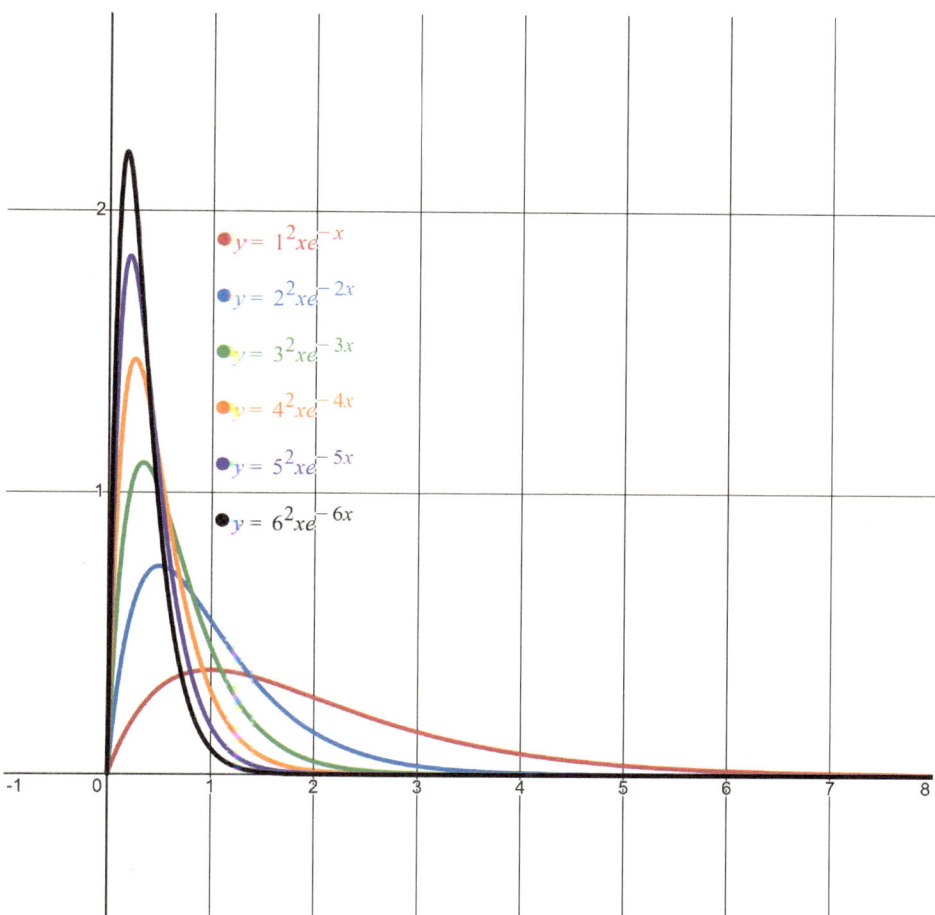

Fig. 6.4 $f_n(x) = n^2 xe^{-nx}$. Image created with the Desmos graphing calculator, used with permission from Desmos Studio FBC

First you must convince yourself that $f_n(x)$ tends to the Dirichlet function pointwise in $[0,1]$. Then it should be noted that although $f_n(x)$ is Riemann integrable on $[0,1]$, the pointwise limit of the sequence (which is the Dirichlet function) is not. That is, this is an example in which the pointwise limit of a sequence of Riemann integrable functions is not Riemann integrable.

Another way to think of the Dirichlet function as the limit of a sequence of functions is as follows:

$$f_n(x) = \lim_{m \to \infty} (\cos(n!\pi x))^{2m}.$$

We suggest that the reader ponder why this sequence of functions converges pointwise to the Dirichlet function.

Activity 6.9:
Prove that $f_n(x)$ tends to the Dirichlet function pointwise in $[0,1]$, that $f_n(x)$ is Riemann integrable (What is the value of its integral from 0 to 1?) and that the Dirichlet function is not Riemann integrable.

The examples above show that the Riemann integral and the pointwise limit don't get along very well. In fact, this is a weakness of the Riemann integral. We have also already seen that the point limit of sequences of continuous functions can be discontinuous, and we will see that the point limit of differentiable functions can be non-differentiable without being discontinuous.

However, these things don't happen with the uniform limit. In fact

- we know that the uniform limit of continuous functions is continuous,
- and it turns out that the uniform limit of Riemann integrable functions is Riemann integrable, and uniform limits and integrals can be exchanged, that is, the integral of the uniform limit must be equal to the limit of the sequence of integrals, and that this limit of the sequence of integrals must exist.

Activity 6.10:
Prove the last proposition above.

Suggestions:

- Observe that:

$$|S(f;Q) - S(f;R)| = |S(f;Q) - S(f_n;Q) + S(f_n;Q) - S(f_n;R) + S(f_n;R) - S(f;R)|$$
$$\leq |S(f;Q) - S(f_n;Q)| + |S(f_n;Q) - S(f_n;R)| + |S(f_n;R) - S(f;R)|$$

- Think of Q and R as refinements of a partition P_ϵ.
- The first and third terms can be bounded by uniform convergence.
- The second term can be limited by the integrability of f_n.
- With this inequality and Cauchy's integrability criterion, proceed to demonstrate the integrability of the uniform limit.

- Now we know that the integral exists, but we don't know what its value is. Argue that the integral of the uniform limit must be equal to the limit of the integrals, and that this limit of integrals must exist.

Example: Consider the following function,

$$f(x) = \sum_{n=0}^{\infty} \frac{\{10^n x\}}{10^n}$$

where the function $\{x\}$ represents the distance of the argument from the nearest integer. First, let's observe that the series that defines $f(x)$ converges uniformly on $\mathbb{R}$ (Why?). Now notice that the partial sums associated with that series define continuous functions:

$$f(x) = \sum_{n=0}^{\infty} \frac{\{10^n x\}}{10^n} = \lim_{N\to\infty} \sum_{n=0}^{N} \frac{\{10^n x\}}{10^n} = \lim_{N\to\infty} f_N(x).$$

Therefore, since the $f_N(x)$ are continuous and converge uniformly to $f(x)$ on $\mathbb{R}$, We have that $f(x)$ is continuous. However, we will see that $f(x)$ is not differentiable at any point. Consider $f'(x)$:

$$f'(x) = \lim_{h\to 0} \frac{f(x+h) - f(x)}{h}.$$

We'll show that for the specific sequence $h_m = \pm 10^{-m}$, the above limit, which defines the derivative, does not exist. In fact, $\frac{f(x+h_m)-f(x)}{h_m} = \pm 10^m \sum_{n=0}^{\infty} \frac{\{10^n x \pm 10^{n-m}\}-\{10^n x\}}{10^n}$. Note that if $n > m$, the numerator is zero. Note also that if $n \le m$, the sign of $h_m = \pm 10^{-m}$ can be chosen so that the numerators are $\pm 10^{n-m}$, so that the series is reduced to a finite sum of m terms of the form ± 1. No series whose terms are such can converge. So, the derivative can't exist at any point. Below we show the very wrinkly graph of the tenth partial sum on $[0,1]$ (Fig. 6.5).

Activity 6.11:
Use the Desmos activity in the book's complementary webpage[2] to plot different partial sums of $f(x) = \sum_{n=0}^{\infty} \frac{\{10^n x\}}{10^n}$ on the online graphing tool and zoom in. Notice how your graph is "wrinkling", i.e. that the graph of $f(x)$ is not "smooth", as is the graph of a differentiable function.

---

[2] https://calculustoanalysis.weebly.com/.

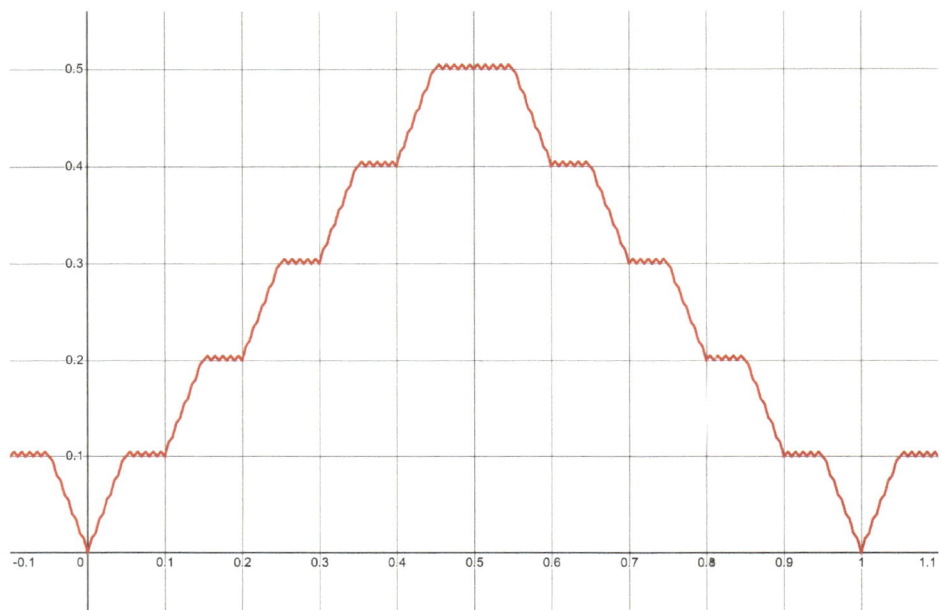

Fig. 6.5 $f(x) = \sum_{n=0}^{10} \frac{\{10^n x\}}{10^n}$. Image created with the Desmos graphing calculator, used with permission from Desmos Studio PBC

Example: Another very famous example of a continuous function that is not differentiable at any point is the following, the so-called Weierstrass function:

$$f(x) = \sum_{n=0}^{\infty} \frac{\cos(3^n x)}{2^n}.$$

It's easy to see that $f(x)$ is continuous, and in fact uniformly continuous (Why?). Now, let's see what happens with

$$\frac{f(x+h) - f(x)}{h} = \sum_{n=0}^{\infty} \frac{\cos(3^n x)(\cos(3^n h) - 1) - \sin(3^n x)\sin(3^n h)}{h 2^n}.$$

If we choose $h = h_n = 2\pi 3^{-n}$, we'll have that

$$\frac{f(x+h_m) - f(x)}{h_m} = \frac{1}{2\pi} \sum_{n=0}^{\infty} \frac{\cos(3^n x)\left(\cos(2\pi 3^{n-m}) - 1\right) - \sin(3^n x)\sin(2\pi 3^{n-m})}{3^{-m}2^n}$$

$$= \frac{3^m}{2\pi} \sum_{n=0}^{m-1} \frac{\cos(3^n x)\left(\cos(2\pi 3^{n-m}) - 1\right) - \sin(3^n x)\sin(2\pi 3^{n-m})}{2^n}$$

$$= \frac{3^m}{2\pi} \sum_{n=0}^{m-1} \frac{\cos(3^n(x + 2\pi 3^{-m})) - \cos(3^n x)}{2^n}$$

.

Note that if $n \geq m$, the numerator is zero. Basically, we would have to show that the resulting finite sum doesn't go zero as $m \to \infty$. In fact, there's a 3^m in the numerator, which is very promising. We suggest that the reader graph the sum for different values of m and x to convince themselves. Perhaps $x = 0$ is an interesting value to start the exploration with. We refer the reader to Körner's book ([4] in Further Readings) for a full demonstration.

Activity 6.12:
Using the grapher, convince yourself that the Weierstrass function has no derivative at any point. Graph $\frac{3^m}{2\pi} \sum_{n=0}^{m-1} \frac{\cos(3^n(x+2\pi 3^{-m})) - \cos(3^n x)}{2^n}$, to convince yourself that it does not tend to zero as $m \to \infty$. In fact, it takes on very large values.

Below are the graphs of the first four partial sums of the Weierstrass function 6.6.

The Weierstrass function is an example of a sequence of uniformly continuous functions (the partial sums), which are also differentiable, which converge uniformly and yet their pointwise limit is nowhere differentiable. In the following figure we show the 11th partial sum of the Weierstrass function. It is already visibly evident that this function is very "wrinkled" everywhere (Fig. 6.7).

Activity 6.13:
Plot different partial sums of $f(x) = \sum_{n=0}^{\infty} \frac{\cos(3^n x)}{2^n}$ in the grapher and notice how the curve is "wrinkling" everywhere, i.e. that the graph of $f(x)$ is not "smooth," as is the graph of a differentiable function. This particular example, discovered by Weierstrass, caused great bewilderment among mathematicians of his day.

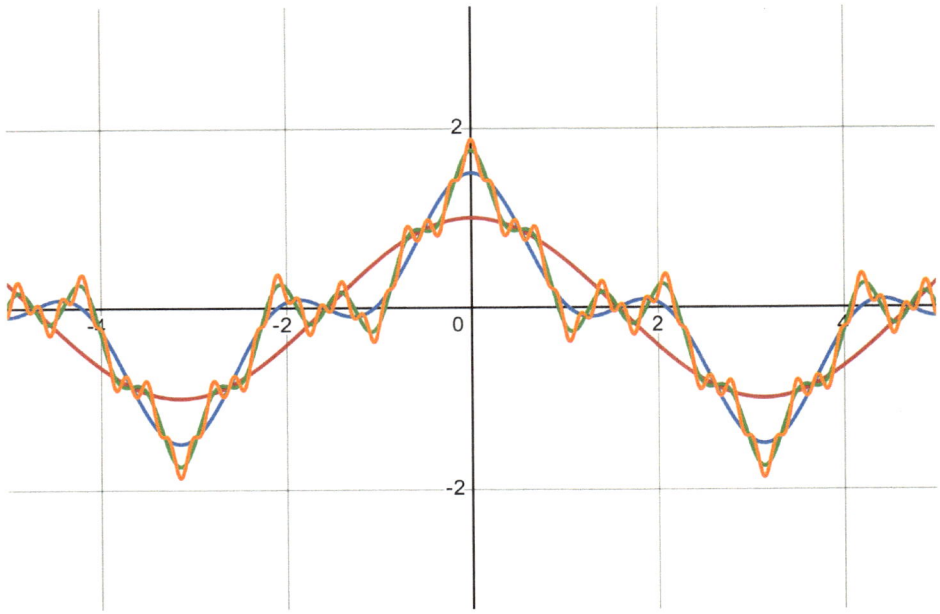

Fig. 6.6   1st four partial sums of the Weierstrass function. Image created with the Desmos graphing calculator, used with permission from Desmos Studio PBC

Let's take one more look at power series representation in the following activity:

Activity 6.14:

Consider the following three functions:

$$f(x) = e^{-x^2}$$
$$g(x) = \frac{1}{1+x^2}$$
$$h(x) = \begin{cases} e^{-1/x^2} & \text{if } x \neq 0 \\ 0 & \text{if } x = 0 \end{cases}$$

Show that (a) $f(x)$ can be represented by a power series centered at $x = 0$ for all x, (b) $g(x)$ can be represented by a power series centered at $x = 0$ for $|x| < 1$, (c) $h(x)$ cannot be represented by a power series centered at $x = 0$.

Note that (c) is very striking, since $h(x)$ is infinitely differentiable at $x = 0$ and that might seem to indicate that a Taylor series would do the job. However, all derivatives at $x = 0$ are 0, and therefore the Taylor series would correspond to the 0 function, which coincides with $h(x)$ only at $x = 0$.

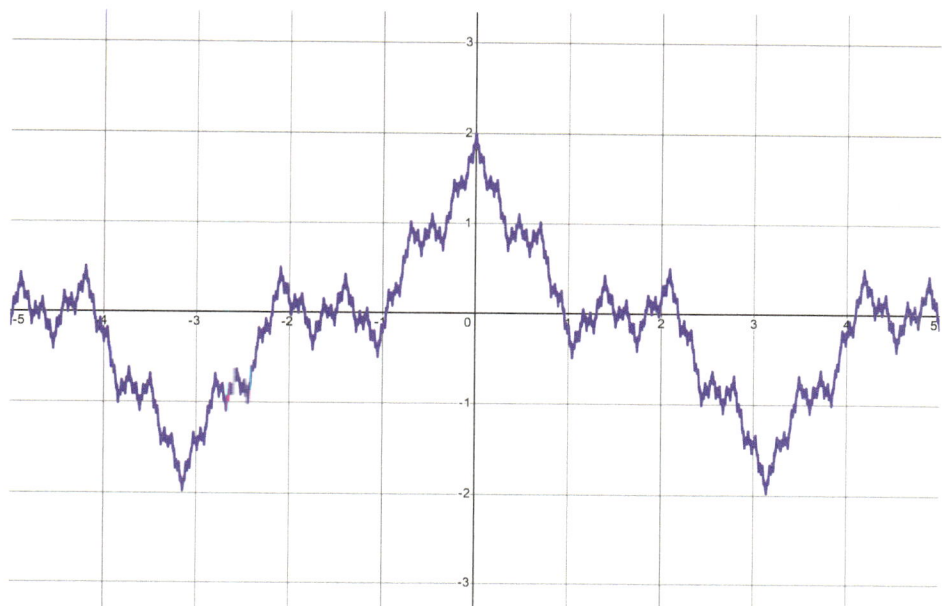

Fig. 6.7 11th partial su π of the Weiestrass function. Image created with the Desmos graphing calculator, used with per mission from Desmos Studio PBC

Note that (b) is also a bit puzzling, since $g(x)$ is well defined for all x and its graph is a "bell curve", not unlike $f(x)$ at first glance. Why then is $f(x)$ representable for all x and $g(x)$ only for $|x| < 1$? The answer lies in the complex plane, where $g(z)$ has poles or infinite discontinuity at $z = i$ and $z = -i$, which are precisely on the unit circle $|z| = 1$, whereas $f(z)$ is well defined for all complex z. This points to the study of analytic functions of a complex variable, a future pursuit to consider.

Are these strange functions exotic, rare specimens, idle pursuits of enthusiasts of the bizarre? As we will see in the next chapters, they are not. These and other bewildering functions arise in the study of trigonometric series, or Fourier series, which Fourier studied in the context of a standard, fundamental application in physics: the study of thermal conduction.

An important chapter in the history of mathematics is the development of the concept of function, which in its most modern, general form, has only been around relatively recently. We can already sense and understand the struggle of going from narrow expressions, like polynomials, which implied continuity and smoothness, to very general rules connecting domain and range, such as we find in the Dirichlet function. Add to that the additional layer of multiple representations for a single function, and the complexity grows. Historically, it was a struggle to convince the mathematical community of the need for the most general definition of function, which is fundamental in mathematical analysis.

Activity 6.15:
In the book's complementary website,[3] you will find links to resources describing the history of the development of the concept of function.

---

[3] https://calculustoanalysis.weebly.com/.

Fourier Series I

7

The non-differentiable Weierstrass continuous function is an example of a trigonometric series. An important part of the study of trigonometric series is known as Fourier series analysis (Fig. 7.1).

We will briefly discuss the historical background of Fourier and his discoveries. Of interest to Fourier and his contemporaries was the study of linear partial differential equations fundamental to mathematical physics, such as the so-called heat equation. This equation governs the behavior of temperature in a bar or beam, for example. We will see how, from the study of these equations, the need to study and understand the behavior of trigonometric series or Fourier series naturally follows.

The heat equation arises by studying the temperature behavior of a bar or beam subject to an initial temperature distribution and to assumed conditions at its extremes or boundaries. If we denote by x the distance of a point on the bar to one end of it, by t time, and by $u(x, t)$ the temperature in a cross-section at a distance x of the reference endpoint at time t, then the differential equation that governs the behavior of this temperature is given by:

$$u_t = ku_{xx},$$

With initial and boundary conditions given by:

$$u(x, 0) = f(x), 0 < x < l$$

$$u(0, t) = u(l, t) = 0,$$

A. Portnoy, *Calculus to Analysis*, Synthesis Lectures on Mathematics & Statistics,
https://doi.org/10.1007/978-3-031-69662-6_7

where $f(x)$ represents the initial temperature of the bar, and the temperatures remain
constant (0) at the ends of the bar. k is a constant that represents the rate of heat diffusion
and depends on the physical properties of the bar.

The fundamental idea to solve this problem is that of separation of variables. Let's
assume that $u(x, t) = X(x)T(t)$. Substituting into the differential equation we get $XT' = kX'T'$ o equivalently $\frac{T'}{kT} = \frac{X'}{X} = \lambda$. We know both quotients must be constant because
one depends only on t and the other one only on x, and they are equal. We call that
constant λ. This results in an eigenvalue problem in X, whose solutions or eigenfunctions
are $X_n(x) = \sin(n\pi x/l)$ with corresponding eigenvalues $\lambda_n = -(n\pi/l)^2$. This results in
the following functions $T_n(t) = e^{k\lambda_n t}$ and finally in solutions $u_n(x, t) = X_n(x)T_n(t) = \sin(n\pi x/l)e^{-k(n\pi/l)^2 t}$. Note that these solutions satisfy the heat equation and boundary
conditions, but not the initial condition. The idea now is to superimpose all these solutions
on a large linear combination, impose the initial condition, and see if we can represent
the initial temperature distribution in terms of this large linear combination:

$$u(x, t) = \sum_{n=1}^{\infty} B_n u_n(x, t) = \sum_{n=1}^{\infty} B_n \sin(n\pi x/l)e^{-k(n\pi/l)^2 t}.$$

Imposing the initial condition:

$$u(x, 0) = \sum_{n=1}^{\infty} u_n(x, 0) = \sum_{n=1}^{\infty} B_n \sin(n\pi x/l) = f(x).$$

The following questions naturally arise:

- Are there coefficients B_n such that the initial condition is satisfied, that is, that the trigonometric series represents the initial condition?
- What conditions must $f(x)$ satisfy for these coefficients to exist?
- If they exist, how do we find them?
- In what sense do we say that the trigonometric series represents the initial condition?

In this section, we will begin to answer only the third question: how do we find the coefficients, assuming that the function is a superposition of trigonometric functions?

Let's assume then that $f(x) = \frac{A_0}{2} + \sum_{n=1}^{\infty} A_n \cos(n\pi x/l) + B_n \sin(n\pi x/l)$. The first thing to notice is that $f(x)$ is periodic with period $2l$. Then we can notice that if we integrate both sides of the equality over the interval $[-l, l]$ we get that $\int_{-l}^{l} f(x)dx = A_0 l$, or equivalently $A_0 = \frac{1}{l} \int_{-l}^{l} f(x)dx$. That is, the first Fourier coefficient is twice the average value of $f(x)$ over a period.

It turns out that these trigonometric functions are called the Fourier basis, and they satisfy the following orthogonality properties:

$$\int_{-l}^{l} \cos(n\pi x/l)\cos(m\pi x/l)dx = \int_{-l}^{l} \sin(n\pi x/l)\sin(m\pi x/l)dx = 0$$

if $n \neq m$, $n, m \in \mathbb{Z}$, and

$$\int_{-l}^{l} \cos(n\pi x/l)\sin(m\pi x/l)dx = 0$$

for all $n, m \in \mathbb{Z}$. Also, we have that if $n = m$:

$$\int_{-l}^{l} \cos^2(n\pi x/l)dx = \int_{-l}^{l} \sin^2(n\pi x/l)dx = l.$$

With these properties it is possible to deduce that:

$$A_n = \frac{1}{l} \int_{-l}^{l} f(x) \cos\left(n\pi x/l\right)$$

$$B_n = \frac{1}{l} \int_{-l}^{l} f(x) \sin\left(n\pi x/l\right)$$

Activity 7.1:
Prove the orthogonality properties of the Fourier basis.

Activity 7.2:
Prove the validity of the formulas for Fourier coefficients using the orthogonality
properties.

Now, using these formulas, we're going to find the first Fourier coefficients of some
functions, construct the Fourier partial sums, and plot them next to the function, to see
what relationship there is between them. Let's forget for a moment that we don't know if
these functions are representable as a Fourier series.

Example: Consider the function $f(x) = \begin{cases} -1 & -\pi < x < 0 \\ 1 & 0 \le x < \pi \end{cases}$. Then the Fourier series
associated with this function is:

$$\frac{4}{\pi}\left[\frac{\sin(x)}{1} + \frac{\sin(3x)}{3} + \frac{\sin(5x)}{5} + \cdots\right].$$

The reader must show that this is so. Now, let's plot some of the partial sums (Fig. 7.2).
Let's make a few observations:

- First, notice that since our function is odd, the Fourier series contains only sines, and
  that the Fourier series "extends" our function periodically.
- The graphs of the partial sums show that the partial sums of the Fourier series seem
  to approximate our function better and better, except perhaps at the points of discontinuity of the function, where we see oscillations deviating from the limit function.
  This oscillatory phenomenon near discontinuities is known as the Gibbs phenomenon.
  Notice that at these jumps or points of discontinuity, the Fourier partial sums all go
  through the average between the one-sided limits of the discontinuous function, which
  is zero in this case.

Activity 7.3:
We suggest that the reader go to the online graphing tool and reproduce these
graphs. Play with larger partial sums to investigate whether convergence is
happening in some sense.

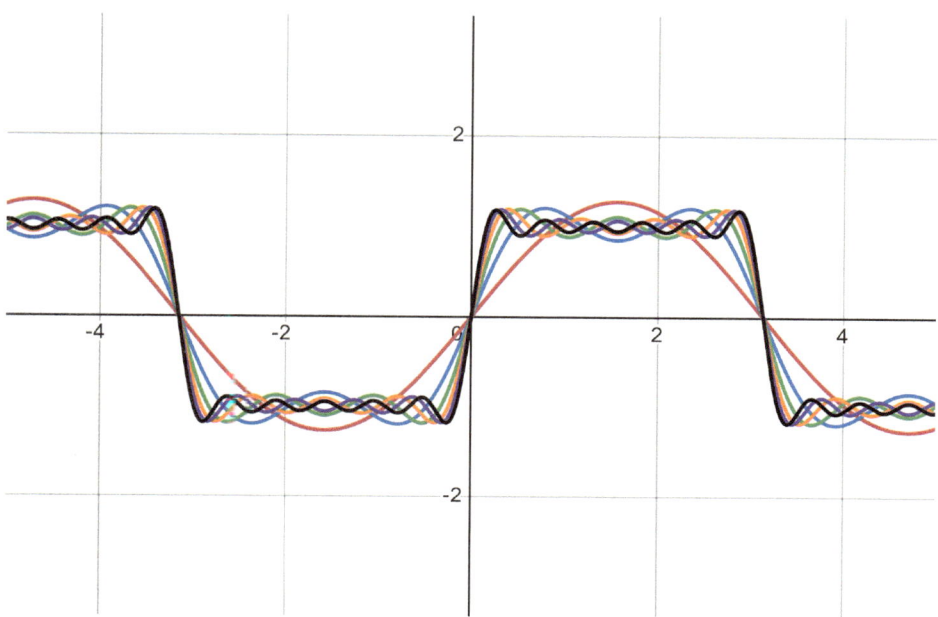

Fig. 7.2 First partial sums of Fourier series. Image created with the Desmos graphing calculator, used with permission from Desmos Studio PBC

Example: Consider the function $f(x) = |x|$ on $[-\pi, \pi]$. Then the Fourier series associated with this function is:

$$\frac{\pi}{2} - \frac{4}{\pi}\left[\frac{\cos(x)}{1^2} + \frac{\cos(3x)}{3^2} + \frac{\cos(5x)}{5^2} + \cdots\right].$$

The reader must show that this is so. Now, let's plot some of the partial sums (Fig. 7.3). Let's make a few pertinent observations:

- First, notice that since our function is even, the Fourier series contains only cosines, and that the Fourier sum "extends" our function periodically.
- The graphs of the partial sums show that the partial sums of the Fourier series seem to approximate our function better and better. In fact, the convergence (informally speaking, visually) of the series seems to be much faster in this example than in the previous one. In fact, the denominator in the nth term of this series shows that it converges absolutely and uniformly. This is not the case in the previous example. In fact, in the first example, the coefficients are proportional to $\frac{1}{n}$, whereas in the second example they are proportional to $\frac{1}{n^2}$.

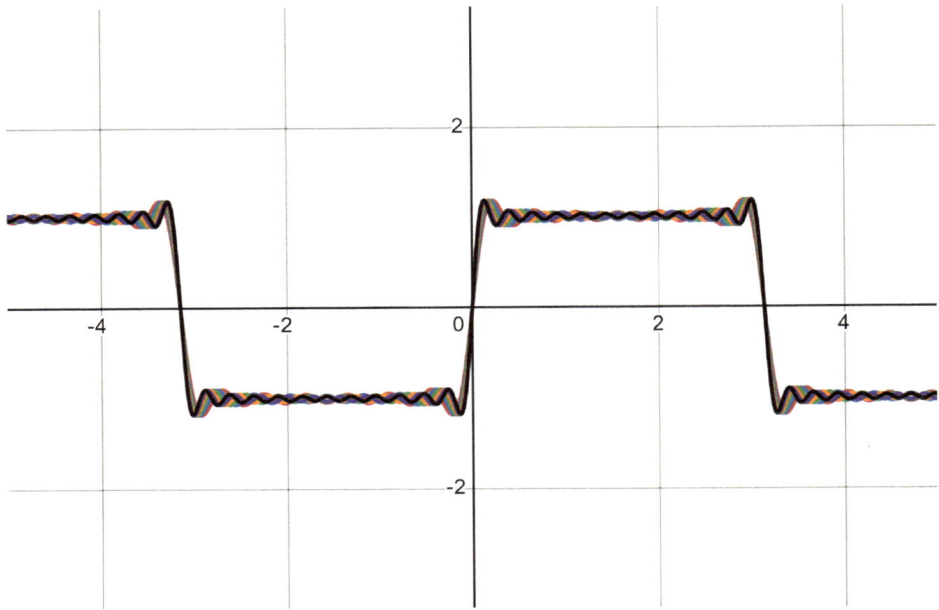

Fig. 7.3 More partial sums of a Fourier series. Image created with the Desmos graphing calculator, used with permission from Desmos Studio PBC

Activity 7.4:
We suggest that the reader go to the online graphing tool and reproduce these graphs. Also, play with larger partial sums, to investigate the convergence of the series.

Activity 7.5:
Let's go back to the Weierstrass example from the previous section:$f(x) = \sum_{n=0}^{\infty} \frac{\cos(3^n x)}{2^n}$, which is a Fourier series. Now let's consider this modified version: $f(x) = \sum_{n=0}^{\infty} \frac{\cos(A^n x)}{B^n}$ where $A, B \in \mathbb{R}$. How should A, B be for the series to converge? How should A, B be for the function to be continuous? How should A, B be to make the function differentiable? How should A, B be so that the function is continuous but not differentiable? We recommend that this activity begin with extensive exploration in the online graphing tool.

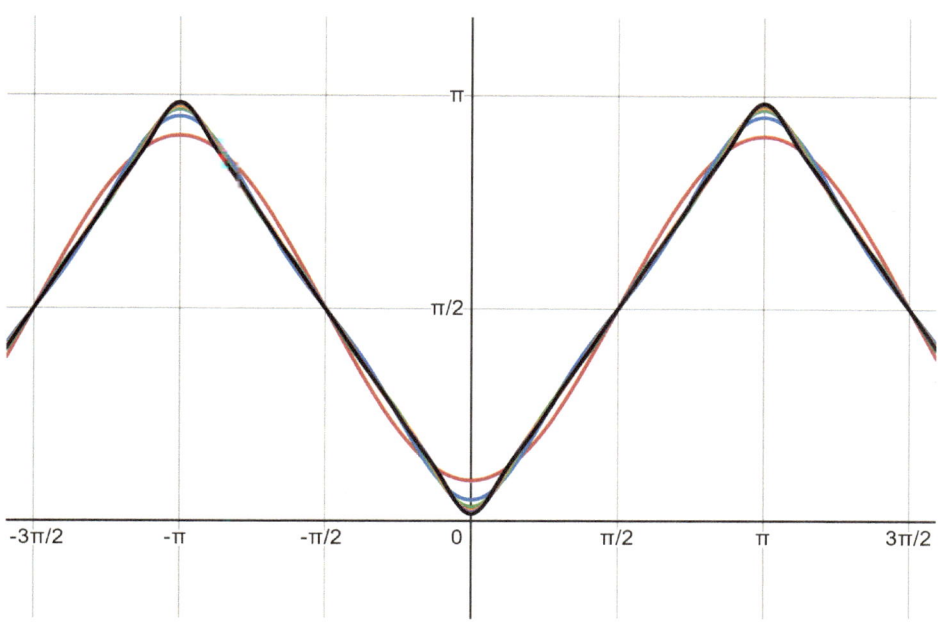

Fig. 7.4 Partial sums of another Fourier series. Image created with the Desmos graphing calculator, used with permission from Desmos Studio PBC

We have already motivated the study of trigonometric series. These arise naturally from the study of differential equations.

Now let's justify the orthogonality and basis terminology used to describe the collection of trigonometric functions and their properties.

Let's consider the set of piecewise continuous functions $f(x) : [a, b] \to \mathbb{R}$ (continuous on $[a, b]$ except for a finite number of jumps or places where one-sided limits exist but do not coincide). Providing the set with the usual sum of functions and with a scalar product on $\mathbb{R}$, we have a vector space; let's denote it by $PC[a, b]$. We can define an inner product for this space as follows:

$$\langle f, g \rangle = \int_a^b f(x)g(x)dx.$$

This inner product induces a norm, $\|f\|^2 = \langle f, f \rangle$, and this norm induces a metric on the vector space:

$$d(f, g) = \|f - g\|.$$

Activity 8.1:
Prove that this defines an inner product, a norm, and a metric in $PC[a, b]$. Look up the definitions of inner product, norm, and metric.

© The Author(s), under exclusive license to Springer Nature Switzerland AG 2025   83
A. Portnoy, *Calculus to Analysis*, Synthesis Lectures on Mathematics & Statistics,
https://doi.org/10.1007/978-3-031-69662-6_8

To facilitate the discussion going forward, we will consider functions in a canonical interval $[-\pi, \pi]$ and the associated Fourier basis:

$$\{1, \sin(x), \cos(x), \sin(2x), \cos(2x), ...\}.$$

If $f(x) \in PC[-\pi, \pi]$, its Fourier coefficients are given by:

$$A_n = \frac{1}{\pi} \int_{-\pi}^{\pi} f(x)\cos(nx)dx = \frac{\langle f(x), \cos(nx) \rangle}{\langle \cos(nx), \cos(nx) \rangle}$$

$$B_n = \frac{1}{\pi} \int_{-\pi}^{\pi} f(x)\sin(nx)dx = \frac{\langle f(x), \sin(nx) \rangle}{\langle \sin(nx), \sin(nx) \rangle},$$

where the orthogonality conditions are:

$$\langle \cos(nx), \cos(mx) \rangle = \int_{-\pi}^{\pi} \cos(nx)\cos(mx)dx$$

$$= \int_{-\pi}^{\pi} \sin(nx)\sin(mx)dx = \langle \sin(nx), \sin(mx) \rangle = 0$$

if $n \neq m$, and

$$\langle \cos(nx), \sin(mx) \rangle = \int_{-\pi}^{\pi} \cos(nx)\sin(mx)dx = 0$$

for all $n, m \in \mathbb{N}$. Also, we have that if $n = m$:

$$\|\cos(nx)\|^2 = \int_{-\pi}^{\pi} \cos^2(nx)dx = \|\sin(nx)\|^2 = \int_{-\pi}^{\pi} \sin^2(nx)dx = \pi.$$

Note that $\|\cos(0x)\|^2 = \|1\|^2 = \int_{-\pi}^{\pi} 1^2 dx = 2\pi$.

We can interpret these coefficients as projections of $f(x)$ on the corresponding elements of the Fourier basis. This basis forms an orthogonal set relative to the inner product. The inner product is what motivates the orthogonality terminology.

Given a function $f(x) \in C[-\pi, \pi]$, we can construct a Fourier series associated with $f(x)$:

$$\frac{A_0}{2} + \sum_{n=1}^{\infty} A_n\cos(nx) + B_n\sin(nx)$$

Now, the questions of interest are:

- Does this series converge?
- In what sense does it converge?
- What does it converge to?

Let us denote the nth partial sum as $f_N(x) = \frac{A_0}{2} + \sum_{n=1}^{N} A_n \cos(nx) + B_n \sin(nx)$. This partial sum is known as the least-squares approximation of $f(x)$ over the subspace generated by $\{1, \cos(x), \sin(x), \cos(2x), \sin(2x), ..., \cos(Nx), \sin(Nx)\}$. This is because of all the elements of that subspace, $f_N(x)$ minimizes $d(f_N(x), f(x))$.

Activity 8.2:
Prove that of all the elements in the subspace, $f_N(x)$ minimizes $d(f_N(x), f(x))$, that is, $f_N(x)$ is the least squares approximation of $f(x)$.

Consider

$$
\begin{aligned}
d^2(f_N(x) + g_N(x), f(x)) &= \|f(x) - f_N(x) - g_N(x)\|^2 \\
&= \langle f(x) - f_N(x) - g_N(x), f(x) - f_N(x) - g_N(x) \rangle \\
&= \langle f(x) - f_N(x), f(x) - f_N(x) \rangle \\
&\quad - 2\langle f(x) - f_N(x), g_N(x) \rangle + \langle g_N(x), g_N(x) \rangle
\end{aligned}
$$

where $g_N(x)$ is an arbitrary element of the same subspace and prove that $\langle f(x) - f_N(x), g_N(x) \rangle = 0$.

In much the same way, we can show that $\pi \left[\frac{A_0^2}{2} + \sum_{n=1}^{N} A_n^2 + B_n^2 \right] = \|f_N(x)\|^2 \leq \|f(x)\|^2 = \int_{-\pi}^{\pi} f^2(x)dx$. This inequality is known as Bessel's inequality.

Activity 8.3:
Prove Bessel's inequality.

Consider

$$
\|f(x) - f_N(x)\|^2 = \langle f(x) - f_N(x), f(x) - f_N(x) \rangle = \langle f(x), f(x) \rangle - 2\langle f(x), f_N(x) \rangle + \langle f_N(x), f_N(x) \rangle
$$

and prove that $\langle f(x), f_N(x) \rangle = \langle f_N(x), f_N(x) \rangle$.
Note that once proven, it implies that

$$
\|f(x) - f_N(x)\|^2 + \|f_N(x)\|^2 = \|f(x)\|^2,
$$

which is just a fancy version of the Pythagorean theorem, which makes sense, given that $f_N(x)$ is the orthogonal projection of $f(x)$ onto the subspace of Fourier basis functions included in the partial sum of the Fourier series, or the least squares approximation (Fig. 8.1).

Fig. 8.1 Bessel inequality. The red vector represents $f(x)$, the green vector represents $f_N(x)$, and the orange vector represents their difference $f(x) - f_N(x)$. Image created with the Desmos graphing calculator, used with permission from Desmos Studio PBC

Note: From Bessel's inequality we can conclude that $\lim_{n \to \infty} A_n = \lim_{n \to \infty} B_n = 0$. This result is called the Riemann-Lebesgue lemma.

From the proof of Bessel's inequality, we can see that:

$$\lim_{N \to \infty} \|f(x) - f_N(x)\|^2 = \lim_{N \to \infty} \langle f(x), f(x) \rangle - \langle f_N(x), f_N(x) \rangle = \|f(x)\|^2 - \lim_{N \to \infty} \|f_N(x)\|^2,$$

So that $\lim_{N \to \infty} \|f(x) - f_N(x)\|^2 = 0$, which would mean $f_N(x)$ tends to $f(x)$ in the sense of the metric $d(\_, \_)$ of mean-square convergence, if and only if $\|f(x)\|^2 = \int_{-\pi}^{\pi} f^2(x)dx = \pi \left[\frac{A_0^2}{2} + \sum_{n=1}^{\infty} A_n^2 + B_n^2 \right] = \lim_{N \to \infty} \|f_N(x)\|^2$. This is known as Parseval's identity.

Activity 8.4:
The purpose of this activity is to investigate the meaning of mean-square convergence.

- Consider the sequence of functions $\{x^n\}$ on the interval $[0,1]$. To what does this succession converge pointwise? Is convergence uniform? Does mean-square convergence occur?
- Consider the sequence of functions $\{nxe^{-nx}\}$ on the interval $[0,1]$. To what does this sequence converge pointwise? Is convergence uniform? Does mean-square convergence occur?
- Does pointwise convergence imply uniform convergence? Does point convergence imply mean square convergence? Does mean-square convergence imply uniform convergence?

Parseval's identity gives us a criterion for whether a function is representable as its Fourier series in the mean-square sense. If we can represent all functions in a vector space in the mean-square sense with respect to the Fourier basis, this justifies the basis terminology. This is our next goal.

Let's denote the set of continuous functions on $[a, b]$ as $C[a, b]$. Let $f(x) \in C[-\pi, \pi]$. Then we have that

$$f_N(x) = \frac{1}{2\pi} \int_{-\pi}^{\pi} f(x)ds + \frac{1}{\pi} \sum_{n=1}^{N} \left[\int_{-\pi}^{\pi} f(x) \cos(ns)ds \cos(nx) + \int_{-\pi}^{\pi} f(x) \sin(ns)ds \sin(nx) \right]$$

$$= \frac{1}{\pi} \int_{-\pi}^{\pi} f(x+s)D_N(s)ds$$

where

$$D_n(s) = \begin{cases} \frac{\sin((n+1/2)s)}{2\sin(s/2)} & 0 < |s| \le \pi \\ n + 1/2 & s = 0 \end{cases}.$$

$D_N(s)$ is called the Dirichlet kernel.

Activity 8.5:
Prove that $f_N(x) = \frac{1}{\pi}\int_{-\pi}^{\pi} f(x+s)D_N(s)ds$. For the proof, it is useful to remember that $e^{ix} = \cos(x) + i\sin(x)$.

Using the Dirichlet kernel, can you prove that if $f(x) = \begin{cases} -1 & -\pi < x < 0 \\ 1 & 0 \le x < \pi \end{cases}$ then $f_N(0)$ is equal to the average of the one-sided limits of $f(x)$ (0 in this case)? Do you think that, in general, the Fourier series of a function with a jump discontinuity at a point converges to the average of the one-sided limits of the function at that point?

Activity 8.6:
Plot the Dirichlet kernel and compare its graph with the graph of the polynomials used in the proof of Weierstrass's theorem: $Q_n(x) = c_n(1 - x^2)^n$. You'll see that the Dirichlet kernel doesn't concentrate its area at the origin like Weierstrass polynomials do. That is, with this kernel we will not be able to demonstrate uniform convergence to $f(x)$.

The following is the Dirichlet kernel graph for some values of n (Fig. 8.2).

However, we will consider a sequence of trigonometric series with improved convergence properties: we will use Cesàro's idea of convergence.

Let

$$\sigma_N(x) = \frac{1}{N}\sum_{n=0}^{N-1} f_n(x) = \frac{1}{\pi}\int_{-\pi}^{\pi} f(x+s)\frac{1}{N}\sum_{n=0}^{N-1} D_n(s)ds$$

$$= \frac{1}{\pi}\int_{-\pi}^{\pi} f(x+s)F_N(s)ds$$

where $F_N(s) = \frac{1}{N}\sum_{n=0}^{N-1} D_n(s) = \begin{cases} \frac{1}{2n}\left(\frac{\sin(ns/2)}{\sin(s/2)}\right)^2 & 0 < |s| \le \pi \\ n/2 & s = 0 \end{cases}$ is called the Fejer kernel.

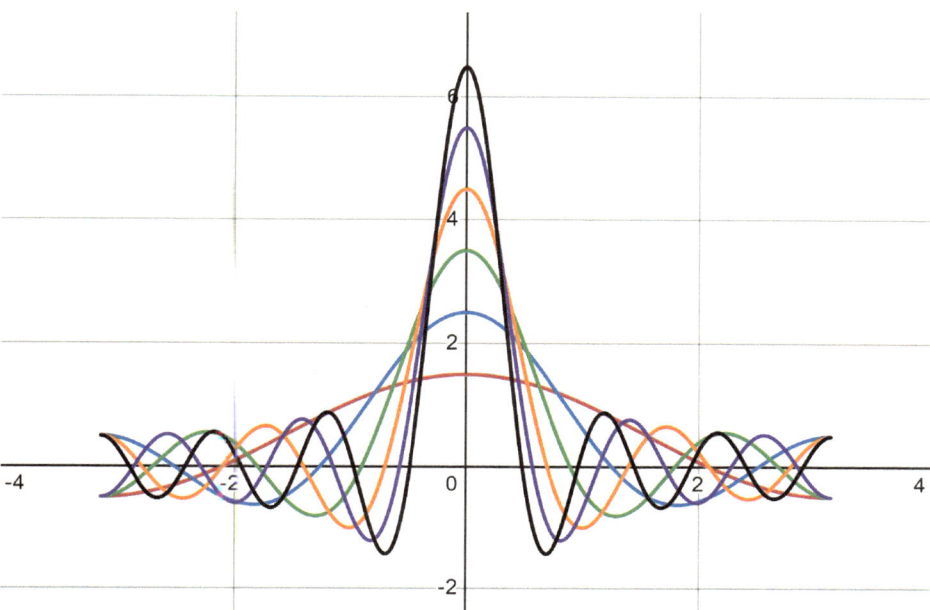

Fig. 8.2 Dirichlet kernel. Image created with the Desmos graphing calculator, used with permission from Desmos Studio PBC

Activity 8.7:

Prove that $F_N(s) = \begin{cases} \frac{1}{2n}\left(\frac{\sin(ns/2)}{\sin(s/2)}\right)^2 & 0 < |s| \leq \pi \\ n/2 & s = 0 \end{cases}$.

Activity 8.8:

Graph the Fejer kernel and compare its graph with the graph of the polynomials used in the proof of Weierstrass's theorem: $Q_n(x) = c_n(1 - x^2)^n$. You'll see that they both concentrate their area at the source. So, using a proof very similar to that of Weierstrass's theorem, we can show that $\sigma_n(x) \underset{unif}{\to} f(x)$. These $\sigma_n(x)$ they are the averages of Fourier partial sums, and they are the sums used to prove Cesàro convergence. The conclusion we can draw from this is that there is a sequence of trigonometric series that converges to $f(x)$ uniformly: $\{\sigma_n(x)\}$.

Below are the Fejer kernel graphs for some values of n (Fig. 8.3).

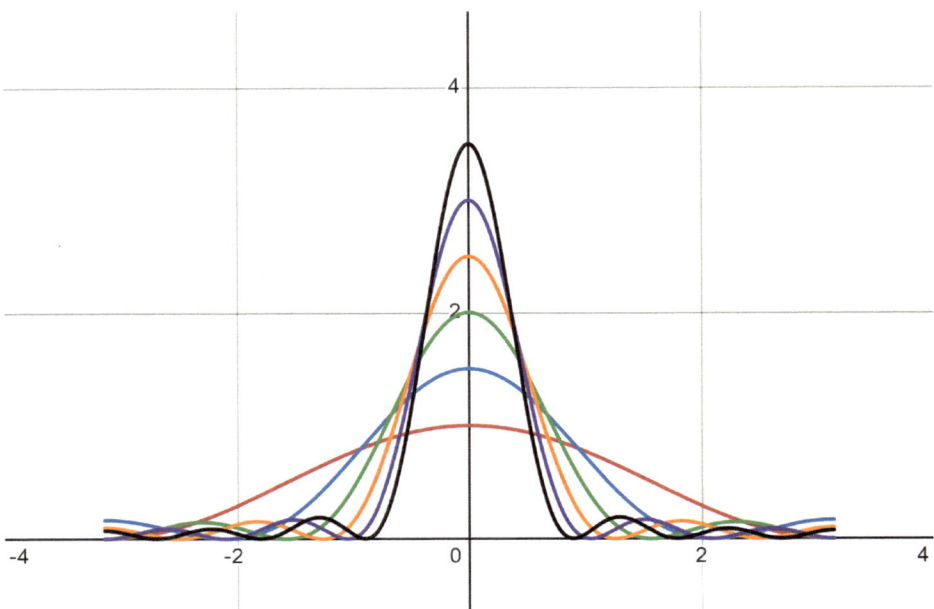

Fig. 8.3 Fejer kernel. Image created with the Desmos graphing calculator, used with permission from Desmos Studio PBC

Now let's go for the big conclusion: Let's show that the Fourier series of any function in $C[-\pi, \pi]$ tends to that function with respect to the mean-square metric. First, let's note that $\|f(x) - f_N(x)\| \leq \|f(x) - \sigma_N(x)\|$, since $f_N(x)$ is the least-squares approximation. We also know that $\|f(x) - \sigma_N(x)\| \leq 2\pi \max_{x \in [-\pi,\pi]} |f(x) - \sigma_N(x)|$. But,

$$\|f(x) - f_N(x)\| \leq \|f(x) - \sigma_N(x)\| \leq 2\pi \max_{x \in [-\pi,\pi]} |f(x) - \sigma_N(x)| \to 0$$

as $N \to \infty$, since $\sigma_N(x) \underset{unif}{\to} f(x)$.

So again, in conclusion, the Fourier series of any function in $C[-\pi, \pi]$ converges to that function with respect to the mean-square metric.

Conclusions and Future Directions

9

We already know that the Fourier series of a continuous function converges to the function in the mean-square sense.

But the following questions immediately arise:

- Is convergence also pointwise?
- What if the function is piecewise continuous?

The Dirichlet-Jordan test states that if a periodic function is of bounded variation (which includes many piecewise continuous functions) then its Fourier series converges at each point in its domain to

$$\lim_{\epsilon \to 0} \frac{f(x - \epsilon) + f(x + \epsilon)}{2},$$

which implies that if f is continuous at x, then the Fourier series converges to $f(x)$ and if there is a jump discontinuity at x, then the Fourier series converges to the average of the one-sided limits of f at x. An example of this last behavior was discussed in the previous chapter.

What is a function of bounded variation? If P is any partition of the interval I, then the total variation of function $f(x)$ defined over I is

$$V_I(f) = \sup_P \sum_{i=1}^{n} |f(x_{i+1}) - f(x_i)|,$$

© The Author(s), under exclusive license to Springer Nature Switzerland AG 2025
A. Portnoy, *Calculus to Analysis*, Synthesis Lectures on Mathematics & Statistics,
https://doi.org/10.1007/978-3-031-69662-6_9

where the supremum is taken over all partitions of the interval I. If $V_I(f)$ is finite, then f is of bounded variation. Functions of bounded variation on a compact interval can also be defined as those which can be written as a difference of bounded monotone functions. A bounded variation function may have discontinuities, but at most countably many.

If f is differentiable and its derivative is Riemann integrable, its total variation is the vertical component of the arc-length of its graph, that is to say,

$$V_I(f) = \int_a^b |f'(x)| dx,$$

where $I = [a, b]$.

Activity 9.1:

(a) Is the Dirichlet function on [0,1] of bounded variation?
(b) Is $f(x) = \sin(1/x)$ of bounded variation on $(0,1]$?
(c) Is $f(x) = x\sin(1/x)$ of bounded variation on $[\epsilon, 1]$ with $0 < \epsilon < 1$?
(d) Is $f(x) = x\sin(1/x)$ of bounded variation on $[0,1]$ (extend f continuously to be defined it at $x = 0$)?
(e) Is a Lipschitz function over $I = [a, b]$ of bounded variation?

It's important to note that the vector space $C[-\pi, \pi]$ is not *complete* with respect to the mean-square metric. In other words, there are Cauchy sequences (in the mean-square sense) in $C[-\pi, \pi]$ whose limit is not in $C[-\pi, \pi]$.

Activity 9.2:
Find a Cauchy sequence, in the mean-square sense, in $C[-\pi, \pi]$ whose limit, in the mean square sense, is not in $C[-\pi, \pi]$.

Activity 9.3:
Find a Cauchy sequence, in the mean-square sense, in $C[-\pi, \pi]$ whose pointwise limit is not in $C[-\pi, \pi]$, but whose limit in the mean-square sense is.

Note that with respect to the mean-square metric, two functions that differ at a single point, or even at a countable collection of points, are "equal" (in mean-square sense).

In fact, the way to make the $C[-\pi, \pi]$ space complete is to endow it with the uniform metric, that is, the metric defined by

$$d(f, g) = \sup_{x \in [-\pi, \pi]} |f(x) - g(x)|.$$

This is the metric of uniform convergence.

Activity 9.4:
Prove that $C[-\pi, \pi]$ endowed with the uniform metric is complete, that is, that any Cauchy sequence, with respect to the uniform metric, in $C[-\pi, \pi]$ converges in $C[-\pi, \pi]$.

Another interesting question is the following: Can a sequence of continuous functions converge pointwise to a function that is not Riemann integrable? The following activity explores this question further.

Activity 9.5:
Can you think of a sequence of continuous functions that converge pointwise to the Dirichlet function? You may remember that another way to think of the Dirichlet function as the limit of a sequence of functions is as follows:

$$f_n(x) = \lim_{m \to \infty} (\cos(n!\pi x))^{2m}$$

However, the sequence $\{f_n(x)\}$ is not a sequence of continuous functions and is therefore not the sequence we are looking for.

So, it turns out that the Riemann integral is not the most appropriate for defining convergence in the mean-square sense. Also, $C[-\pi, \pi]$ isn't complete under that metric.

These reasons and the impossibility of ensuring valid exchanges between integrals and point limits with the Riemann integral largely prompted the development of measure theory and a new definition of integral that generalizes the Riemann integral and defines a natural space of functions in which Fourier analysis can take place. This new integral is the Lebesgue integral. The following is a result, the demonstration of which requires the techniques and concepts developed by Lebesgue, but which applies to the Riemann integral.

Bounded or dominated convergence theorem: Let $\{f_n\}$ be a sequence of functions that are Riemann integrable on $[a, b]$. Suppose there is a positive constant $B > 0$ such that $|f_n(x)| < B$ for all $n \in \mathbb{N}, x \in [a, b]$. If the pointwise limit of the sequence exists and is Riemann integrable on $[a, b]$, then $\int_a^b \lim_{n \to \infty} f_n(x)dx = \lim_{n \to \infty} \int_a^b f_n(x)dx$.

This theorem allows integrals and limits to be exchanged under much more relaxed conditions, but it has the disadvantage that it must assume that the limit of the dominated or bounded sequence is Riemann integrable. To get rid of this assumption one must switch to the Lebesgue integral. The example to keep in mind is the Dirichlet function.

Curious examples of functions have been presented throughout the book. Some may still seem exotic, unusual, even capricious. However, the modern concept of a function was historically difficult for mathematicians to accept and is an important chapter in the development of mathematical analysis. These examples paved the way. As Fourier series were introduced it became increasingly clear that seemingly pathological examples of functions arise quite naturally from the study of trigonometric series and the functions they represent. Given that Fourier was inspired by a fundamental application in physics, heat conduction, these exotic functions represented by trigonometric series, become even more relevant.

We shall conclude this discussion by stating a series of very interesting, true propositions without proof, which we hope will motivate the curious student to study, among other things, measure theory and the Lebesgue integral. These propositions represent some convergence properties of Fourier series; their proofs occupied great mathematicians for more than a century:

• The Fourier series of functions whose square is Lebesgue integrable converge to those functions in the mean-square sense (using the Lebesgue integral).
• There are Lebesgue integrable functions, which are not Riemann integrable, whose Fourier series diverge at every point. (Kolmogorov)
• For any continuous function, its Fourier series converges pointwise to the function except on a set of measure 0, that may be empty. (Carleson)
• Given a set of measure 0, there exists a continuous function whose Fourier series diverges on all points in that set. (Kahane & Katznelson)
• Continuous functions exist whose Fourier series diverge on an infinite non-countable set. (Kahane & Katznelson)

In Appendix III there is a brief discussion about Lebesgue measure and integration.

Appendix A: A Bit on the History of Calculus

This appendix contains a few landmarks in the history of Calculus.

As stated previously, the problem of calculating the area of general plane regions is ancient and may have its origins on the land surveying problem, which arises from the advent of agriculture, thousands of years ago. Farmers had to pay tribute to their overlords, and this tribute was proportional to the area of land under their care. Landowners left their land to their heirs, dividing it required calculating areas of irregular land parcels.

It is curious that in the typical calculus course, the area problem is relegated to the end of the course, since it is related to the integral. Typically, one learns about limits, continuity, differentiability, optimization, and then proceeds to the integral and the problem of area. It makes sense conceptually, after the fact, but it is not how the ideas developed historically. The efficient presentation of mathematical results often does not follow the chronological order in which they were developed. However, it is both interesting and enlightening to consider the chronological order of things, to get a better understanding of the origins of the ideas.

One of the first documented instances of the area problem is known as Dido's problem, which according to legend was named after the founder and first queen of Carthage, Dido (ninth century BC). The problem consisted of enclosing the maximum amount of land bounded by a straight coastline, with the aid of a finite length rope, to define the boundaries of a new coastal city, Carthage. Dido intuitively deduced that the rope should define a semicircular boundary. Thus, the enclosed city, Carthage, would have a semicircular shape. This is an example of the so-called isoperimetric problem, which was not rigorously solved until the nineteenth century.

A. Portnoy, *Calculus to Analysis*, Synthesis Lectures on Mathematics & Statistics,
https://doi.org/10.1007/978-3-031-69662-6

Fig. A.1 Archimedes of
Syracuse (287–212 BC).
Attribution: By Domenico
Fetti—http://archimedes2.
mpiwg-berlin.mpg.de/archim
edes_templates/popup.htm,
Public Domain, https://com
mons.wikimedia.org/w/index.
php?curid=146592

The Method of Exhaustion

Eudoxus (408–355 BC) and Archimedes (287–212 BC) developed and used the method
of exhaustion to calculate areas of regions delimited by certain geometric shapes.
Archimedes used the method of exhaustion to estimate the area of a circle and therefore to estimate the value of π. He inscribed and circumscribed the circle with n sided
regular polygons, calculated their areas thus providing lower and upper estimates for the
area of the circle, and argued what would happen to those estimates as the number of
sides n became larger (Fig. A.1).

Archimedes also used related ideas to find the area under a parabolic segment, a
problem known as the quadrature of the parabola. We will explore this problem in the
following activity.

Activity A1.1:
In this activity, we will follow Archimedes' ideas to calculate the area in a parabolic
segment. A parabolic segment is the area between a parabola and a line secant
to it. Consider the area of the parabolic segment enclosed by the graph of $y =$

$f(x) = x^2$ and the secant line that connects the points $(x_0 - h, f(x_0 - h))$ and $(x_0 + h, f(x_0 + h))$.

- Show that the area of the triangle with vertices at the points $(x_0 - h, f(x_0 - h))$, $(x_0 + h, f(x_0 + h))$ and $(x_0, f(x_0))$ is equal to h^3. Hint: use the fact that the area of a trapezium or right trapezoid is the average of its two parallel sides times the distance between them.
- Now, through a process of bisection, begin exhausting or filling in the parabolic segment with the above triangle and two more triangles like the green ones shown in the illustration below (Fig. A.2).

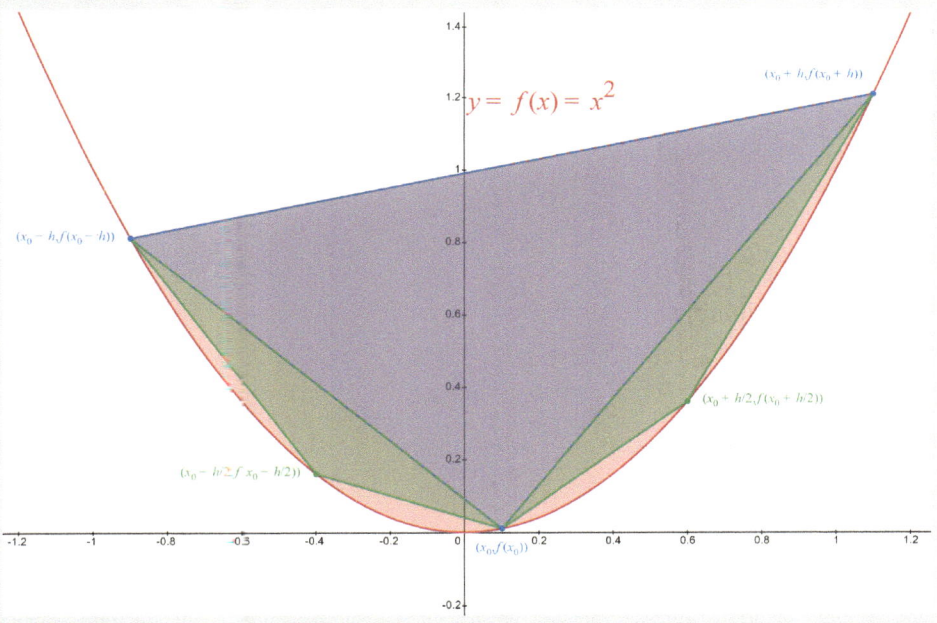

Fig. A.2 Method of exhaustion, quadrature of the parabola. Image created with the Desmos graphing calculator, used with permission from Desmos Studio PBC

- We know that the area of the blue triangle is h^3, and analogously we can conclude that the area of each of the green triangles is $\left(\frac{h}{2}\right)^3 = \frac{h^3}{8}$. Since there are two of them, the green area is $\frac{h^3}{4}$.

- Continuing this exhaustion process we would get that the area of the parabolic segment is $h^3\left(1 + \frac{1}{4} + \frac{1}{16} + \frac{1}{64} + \ldots\right) = \frac{4}{3}h^3$. (Why?)

Cavalieri

A very useful and ingenious principle was discovered by Bonaventura Francesco Cavalieri (1598–1647) which simply states that if two regions in the plane/space can be sliced following parallel lines/planes and slices on the same line/plane have the same length/area, then both regions have the same area/volume. This principle contains the seeds of ideas like exhaustion, infinitesimals, and limits, which are crucial elements of what we now know as calculus (Fig. A.3).

Fig. A.3 Bonaventura Francesco Cavalieri (1598–1647). Attribution: By Unknown author—Trattato della sfera e prattiche per vso di essa, Roma, 1682, Public Domain, https://commons.wikimedia.org/w/index.php?curid=8546341

Activity A1.2:
In this activity, we will use Cavalieri's principle to show that the area in a parabolic segment does not depend on where on the parabola the segment is, but only on the length of the interval on x over which it is defined. Consider the area of the parabolic segment enclosed by the graph of $y = f(x) = x^2$ and the secant line that connects the points $(x_0 - h, f(x_0 - h))$ and $(x_0 + h, f(x_0 + h))$.

- Define the secant line as the graph of the linear function $g(x)$.
- Show that $g(x) - f(x) = h^2 - (x - x_0)^2$.
- Now, use Cavalieri's principle by slicing the parabolic segment in vertical slices, of length $g(x) - f(x)$ and note that this depends only on h and the distance from x_0.
- Conclude that all parabolic segments which correspond to an interval over x of length $2h$ have the same area (Fig. A.4).

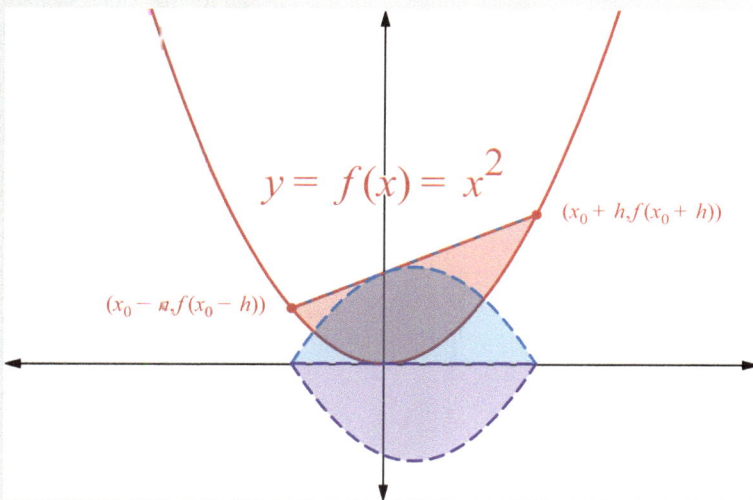

Fig. A.4 Cavalieri's principle. Image created with the Desmos graphing calculator, used with permission from Desmos Studio PBC

You can visit the above interactive demonstration on the book's complementary webpage[1] which shows that the red region representing a parabolic segment occupies the same area as the blue or the purple regions, which represent the area of the parabolic segment centered at the origin. That is, all parabolic segments over an interval of length $2h$ occupy the same area.

Newton and Leibniz

Physics and mathematics were both ripe for the coming together of the ideas of area, slope, distance, instantaneous velocity, limits, derivatives, integrals, etc.; so ripe in fact that two great minds, Newton and Leibniz, formalized and recognized in all these ideas the foundations of the calculus, almost simultaneously, without one knowing the work of the other. The details of this story, and the ensuing rivalries and fights over who was first and the intellectual rights of these ideas, are part of one of the most famous chapters in the history of mathematics (Fig. A.5).

Fig. A.5 Sir Isaac Newton (1642–1726/27). Attribution: By Godfrey Kneller—File: Portrait of Sir Isaac Newton, 1689.jpg from https://exhibitions.lib.cam.ac.uk/linesofthought/artifacts, Public Domain, https://commons.wikimedia.org/w/index.php?curid=132521185

---

[1] https://calculustoanalysis.weebly.com/.

Activity A1.3:

In this activity, we will use Newton's and Leibniz's ideas to calculate the area in a parabolic segment. Consider the area of the parabolic segment enclosed by the graph of $y = f(x) = x^2$ and the secant line that connects the points $(x_0 - h, f(x_0 - h))$ and $(x_0 + h, f(x_0 + h))$.

- Using Cavalieri's principle, which assures us that the area of all parabolic segments with constant h is the same, set up an integral that represents the area of the parabolic segment that corresponds to $x_0 = 0$.
- Calculate the integral using the Fundamental Theorem of Calculus.
- Check that the answer obtained coincides with the answer obtained by Archimedes' exhaustion technique.
- Check that the answer obtained remains invariant to the choice of x_0 (Fig. A.6).

Fig. A.6 Gottfried Wilhelm Leibniz (1646–1716). Attribution: By Christoph Bernhard Francke—Herzog Anton Ulrich-Museum, online, Public Domain, https://commons.wikime dia.org/w/index.php?curid=53159699

Given that distance travelled is time multiplied by velocity, then it becomes clear that for non-constant velocities, net distance traveled corresponds to the area under the velocity curve. In one fell swoop, derivative and integral are related using the physical and geometrical interpretation of these quantities.

Newton's interest in the dynamics of the celestial bodies was related to his interest in calculus. Newton's laws of motion involve the position of an object subject to forces, its instantaneous velocity and acceleration, 1st and 2nd derivatives of the position function. Therefore, these concepts are prerequisites to his theory of dynamics and to being able to formulate these physical laws mathematically.

Activity A1.4:
A great achievement of Newton, once the calculus and his laws of motion, including his law of universal gravitation, had been formulated, was using them to explain Kepler's laws of planetary motion.

Read about the derivation of Kepler's laws of planetary motion from Newton's laws using the links and resources provided in the book's complementary webpage[2] and others you may find on your own.

---

[2] https://calculustoanalysis.weebly.com/.

Appendix B: Heine-Borel Theorem

We defined compact set so that $A \subset \mathbb{R}$ is compact if it is closed and bounded, then remarked that the definition of a compact set in many references is that a set is compact if given any collection of open sets containing the set, we can always find a finite subcollection whose union also contains the set. The equivalence of these two definitions in $\mathbb{R}$ (and $\mathbb{R}^n$) is known as the Heine-Borel theorem, which follows.
For a subset $S \subset \mathbb{R}$, the following two statements are equivalent:

(a) S is closed and bounded.
(b) Every open cover of S has a finite subcover.

What is an open cover of S? It is a collection of open sets whose union covers or contains the set S.

Let's prove this theorem. We will start with $(a) \Rightarrow (b)$. We assume that S is closed and bounded. Since S is bounded, it is contained in the interval $I_0 = [-r, r]$ for some $r > 0$. Now we begin a bisection process where we split the interval I_0 in halves to prove, by contradiction, that every open cover of I_0 admits a finite subcover. Let C be and open cover of I_0 which we assume not to have a finite subcover. We assume C to be infinite; if it were not, C itself would be a finite subcover, and we would have a contradiction. Consider both halves of $I_0 = [-r, r] = [-r, 0] \cup [0, r]$. If each half admits finite subcovers, then we would have a contradiction, since the union of both finite subcovers would be a finite subcover of I_0. So, one of these halves does not admit a finite subcover. Call that half I_1. Now repeat the bisection process and the argument, thus producing a nested sequence of closed intervals $I_0 \supset I_1 \supset I_2 \supset \cdots$ which do not admit a finite subcover and whose length is being halved at each step. The intersection of this sequence of closed, nested intervals, whose length tends to zero, is a point $x \in I_0$ (we observed this when proving the intermediate value theorem). Since this point is covered by some open set $A \in C$, then it is an interior point of the open set A, and therefore, there is an n such that $I_r \subset A$. But this is a contradiction, since I_n does not admit a finite subcover, and A would cover it. We have proven that every open cover of I_0 has or admits a finite subcover.

A. Portnoy, *Calculus to Analysis*, Synthesis Lectures on Mathematics & Statistics, https://doi.org/10.1007/978-3-031-69662-6

Fig. B.1 Nested intervals.
Image created with the Desmos
graphing calculator, used with
permission from Desmos
Studio PBC

Now, let C' be an open cover of S. Note that if we add the open set S^c to C', we have an open cover for I_0. But we now know that it must admit a finite subcover for I_0. If that finite subcover contains S^c, remove it, it is not necessary to cover S. The result is a finite subcover of S drawn from C'. Now we have proven that every open cover of S admits a finite subcover.

Note that we have already encountered this powerful idea of closed, nested intervals in proving the Bolzano-Weierstrass theorem and the intermediate value theorem for continuous functions (Fig. B.1).

Now let's prove $(b) \Rightarrow (a)$. We assume every open cover of S admits a finite subcover. Consider the following open cover of S: let $\epsilon > 0$ and for every $x \in S$ take $(x - \epsilon, x + \epsilon)$ to be one of the cover elements. Now consider the finite subcover which we assume exists and denote it by $\{(x_1 - \epsilon, x_1 + \epsilon), (x_2 - \epsilon, x_2 + \epsilon), \cdots, (x_n - \epsilon, x_n + \epsilon)\}$ where $x_1, x_2, \cdots, x_n \in S$ and $x_1 < x_2 < \cdots < x_n$. Then clearly $S \subset (x_1 - \epsilon, x_n + \epsilon)$ and is therefore bounded. Now, to prove that S is closed, we will prove that its complement S^c is open. Let $y \in S^c$ and now consider the open cover of S such that for every $x \in S$ take $(x - |x - y|/2, x + |x - y|/2)$, to be one of the cover elements. Now consider the finite subcover $\{(x_1 - |x_1 - y|/2, x_1 + |x_1 - y|/2), (x_2 - |x_2 - y|/2, x_2 + |x_2 - y|/2), \cdots, (x_n - |x_n - y|/2, x_n + |x_n - y|/2)\}$ where $x_1, x_2, \cdots, x_n \in S$. Then if $\delta = \min\{|x_1 - y|, |x_2 - y|, \cdots, |x_n - y|\} > 0$ we have that $(y - \delta/2, y + \delta/2) \subset S^c$ since it does not intersect with any of the elements of the finite subcover of S and therefore does not intersect with S. Therefore y is an interior point of S^c, and since this holds for every $y \in S^c$, then S^c is open, making S closed.

Note that this proof can be generalized for sets in $\mathbb{R}^n$.

An observation that is timely at this point:

Any collection of closed sets contained in a compact set such that any intersection of finitely many sets in the collection is non-empty, has a non-empty intersection.

In the Heine-Borel theorem proof we have considered nested, closed intervals whose length tends to zero, which results in an intersection that is a single point.

¿How would one prove the more general observation?

Let $C \subset \mathbb{R}$ be our compact set and let $\{C_i\}$ be our collection of nested closed sets contained in C. Let's proceed by contradiction. Suppose that $\bigcap_i C_i = \varnothing$. The $\{C_i^c\}$ is an open cover for $\mathbb{R}$, and therefore for C as well, and since C is compact, it admits a finite subcover $\{C_{j_i}^c\}_{i=1}^n$, that is $C \subset \bigcup_{i=1}^n C_{j_i}^c$, but this would imply that $\bigcap_{i=1}^n C_{j_i} \subset C^c$, but since each $C_{j_i} \subset C$ then $\bigcap_{i=1}^n C_{j_i} = \varnothing$, which would contradict the assumption that any finite intersection of elements in the collection is non-empty. Note that we are using De Morgan's laws for unions and intersections of sets.

Appendix C: Lebesgue Measure and Integral

The basic idea behind the Riemann integral is to partition the domain into subintervals, grab representative values of the integrand for each subinterval, take the product between subinterval lengths and representative function values, and summing over all subintervals. Finally, the limit is taken as the subintervals become smaller and smaller (Fig. C.1).

In contrast and roughly speaking, the Lebesgue integral can be defined by partitioning the range of the function into subintervals, finding subsets of the domain mapping into each partition ("level sets" defined by inequalities), multiplying a representative value of the function for each subinterval times the Lebesgue measure of the corresponding subset of the domain ("level set") and summing over all subintervals. Finally, the limit is taken as the subintervals become smaller and smaller.

Fig. C.1 Henri Léon Lebesgue (1875–1941). Attribution: Public Domain, https://com mons.wikimedia.org/w/ind ex.php?curid = 336,482

A. Portnoy, *Calculus to Analysis*, Synthesis Lectures on Mathematics & Statistics,
https://doi.org/10.1007/978-3-031-69662-6

Activity C1.1:
In this activity, use the interactive graph in the book's complementary webpage[3] to see the different approaches of the Riemann integral (partitioning the domain) vs Lebesgue's integral (partitioning the range) (Fig. C.2).

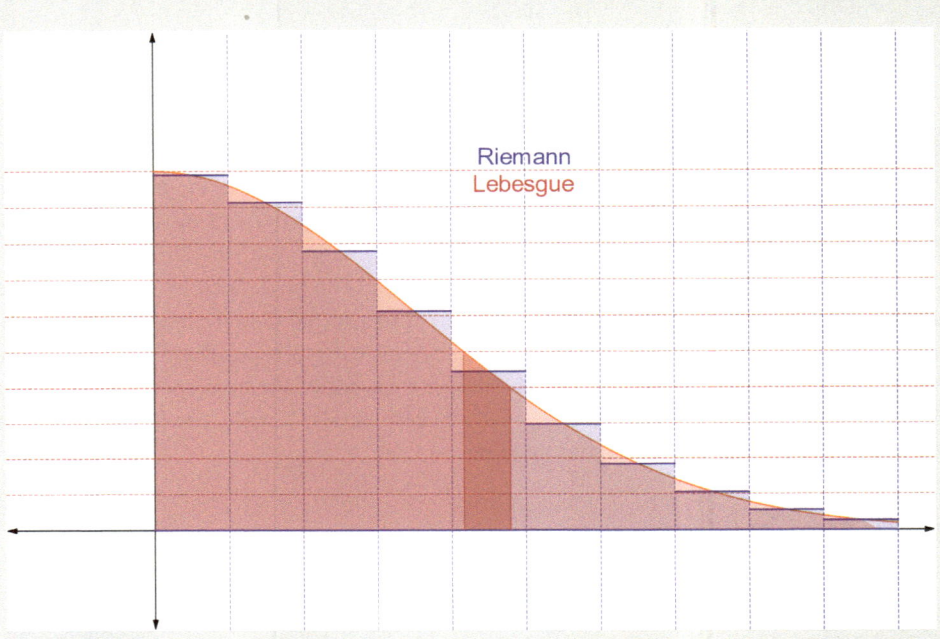

Fig. C.2 Riemann vs. Lebesgue integral. Image created with the Desmos graphing calculator, used with permission from Desmos Studio PBC

This new way of calculating integrals requires defining clearly what the measure of these subsets of the domain of the integrand ("level set") is. These sets can be quite complex, unlike those shown in the previous activity, which are plain intervals. Are all possible subsets measurable? If not, does this impose serious restrictions on the new integral? These questions and many more are answered by Lebesgue measure and integration theory. Lebesgue measurable functions are those whose "level sets" are measurable, making them amenable to Lebesgue integration.

So, what is the measure of a set in $\mathbb{R}$? A measure is a generalization of the notions of length, area, volume that includes very general sets. A measure is a function μ that goes

---

[3] https://calculustoanalysis.weebly.com/.

from a collection of measurable sets to $[0, \infty)$ and must satisfy certain basic properties that length, area and volume satisfy too:

- It should be non-negative.
- The measure of the empty set should be 0.
- It should be countably additive, that is, that the measure of the union of a countable collection of pairwise disjoint measurable sets should equal the sum of their individual measures.

This measure function does not work on any subset of $\mathbb{R}$, but on a collection of measurable subsets that form a σ-algebra. A σ-algebra on a set X is a nonempty collection Σ of subsets of X closed under complement, countable unions, and countable intersections. The ordered pair (X, Σ) is called a measurable space, the members of Σ are called measurable sets. The ordered triple (X, Σ, μ) is called a measure space.

Let's assume our σ-algebra contains all closed intervals and let's define the measure of a closed interval $\mu([a, b]) = b - a$ as its usual length. This will eventually result in the so-called Lebesgue measure in $\mathbb{R}$. Note that the measure of a single point is therefore 0. Consequently, the measure of an open interval $\mu((a, b)) = b - a$ (Why?). We'll use this to figure out the Lebesgue integral of some examples we have studied.

Note that the union of a finite or a countable collection of single points has 0 Lebesgue measure. For example, the set of rational numbers in the interval $[0,1]$, which we already know is countable, has Lebesgue measure of 0. Using the above notation, it would write $\mu(\mathbb{Q} \cap [0,1]) = 0$.

To distinguish Riemann integrals from Lebesgue integrals we will use different notations. For example, the Riemann integral of $f(x)$ over the interval $[a, b]$ we will denote by $\int_a^b f(x)dx$, whereas the Lebesgue integral of $f(x)$ over the interval $[a, b]$ we will denote by $\int_{[a,b]} f(x)dx$.

Activity C1.2:
Riemann and Lebesgue integral of Heaviside function. Remember the *Heaviside step function*:

$$f(x) = \begin{cases} 0 & \text{if } x < 0 \\ 1 & \text{otherwise} \end{cases}.$$

Now consider the Riemann integral from -1 to 1:

$$\int_{-1}^{1} f(x)dx = \int_{-1}^{0} f(x)dx + \int_{0}^{1} f(x)dx = \int_{-1}^{0} 0dx + \int_{0}^{1} 1dx = 1.$$

What about the Lebesgue integral? Well, $f(x)$ takes on only two values, 0 over $[-1,0)$ and 1 over $[0,1]$. So, the Lebesgue integral would be:

$$\int_{[-1,1]} f(x)dx = 0 \cdot \mu([-1,0)) + 1 \cdot \mu([0,1]) = 0 \cdot 1 + 1 \cdot 1 = 1.$$

Both integrals coincide, which is nice. When both integrals exist, it would be nice for them to coincide. Do you think this will this hold in general?

Activity C1.3:
Riemann and Lebesgue integral of the Dirichlet function. Remember Dirichlet's function:

$$f(x) = \begin{cases} 1 \ x \in \mathbb{Q} \\ 0 \ x \notin \mathbb{Q} \end{cases}.$$

Now consider the Riemann integral from -1 to 1. Unfortunately, no matter how we partition the interval $[-1,1]$ in each subinterval we can always choose to sample the function at a rational or at an irrational point. If we always make the first choice, then the integral will result in 2. If we always make the second choice, it will result in 0. What does this imply for the Riemann integral of this function over $[-1,1]$?

What about the Lebesgue integral? Well, $f(x)$ takes on only two values, 0 over $\mathbb{R}\backslash\mathbb{Q}$ and 1 over $\mathbb{Q}$. So, the Lebesgue integral would be:

$$\int_{[-1,1]} f(x)dx = 0 \cdot \mu([-1,1] \cap (\mathbb{R}\backslash\mathbb{Q})) + 1 \cdot \mu([0,1] \cap \mathbb{Q}).$$

What is $\mu([-1,1] \cap (\mathbb{R}\backslash\mathbb{Q}))$? What is $\mu([-1,1] \cap \mathbb{Q})$? What does this imply about the Lebesgue integral of Dirichlet's function over $[-1,1]$? Do both integrals coincide?

Activity C1.4:

Riemann and Lebesgue integral of Cantor's function. Let's remember Cantor's function: Let $f(x) : [0,1] \to [0,1]$ be defined as follows:

- First, express x in ternary.
- If the resulting ternary representation contains the digit 1, replace every ternary digit following the 1 by a 0.
- Now replace all 2's with 1's.
- Finally, interpret the result as a binary number which is then $f(x)$.

As we have mentioned before, this construction produces "steps" in the middle thirds. There is another way of constructing this function as the limit of a sequence of continuous functions as follows (Fig. C.3):

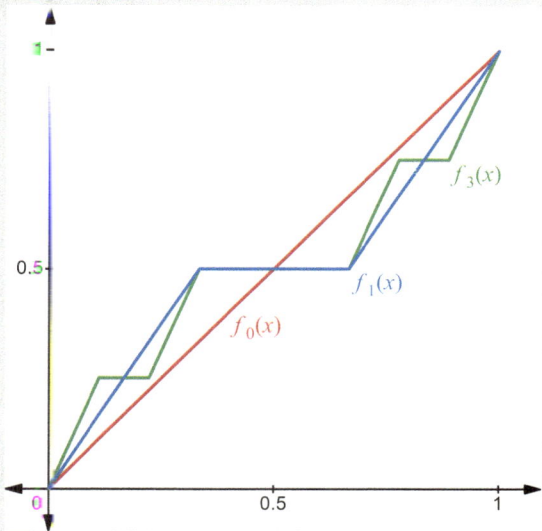

Fig. C.3 Iterative construction of Cantor's function. Image created with the Desmos graphing calculator, used with permission from Desmos Studio PBC

In this iterative construction, we see the first terms of a sequence of piecewise linear functions whose graphs go from (0,0) to (1,1), which are continuous, increasing, and piecewise linear. They connect the "steps" which the previous construction set at each middle third.

(a) Prove that the sequence converges uniformly in [0,1] (Hint: use the Cauchy criterion).
(b) Prove that each sequence element is uniformly continuous.
(c) Conclude that the limit function is continuous as well.
(d) Prove that the area under the graph of each sequence element (its Riemann integral) on [0,1] is equal to1/2.
(e) Conclude that the limit function is Riemann integrable, and its Riemann integral is equal to 1/2 as well. What about its Lebesgue integral? Do you think it exists as well?

Example: (Thomae) Consider Thomae's function:

$$f(x) = \begin{cases} \frac{1}{q} & \text{if } x \in \mathbb{Q} \text{ and } x = \frac{p}{q} \text{ is a reduced fraction} \\ 0 & \text{otherwise} \end{cases}$$

This function is continuous at all irrational numbers and discontinuous at all rational numbers.

Activity C1.5:
Consider Thomae's function, which we know is continuous on all irrational numbers and discontinuous on all rational numbers:

$$f(x) = \begin{cases} 1/q & \text{if } x \in \mathbb{Q} \text{ and } x = p/q \text{ is a reduced fraction} \\ 0 & \text{otherwise} \end{cases}$$

Do you think this function is Riemann integrable? Is it Lebesgue integrable? Which is easier to determine?

This example brings us to Lebesgue's criterion for Riemann integrability, which we state without proof: Let $f(x) : [a, b] \to \mathbb{R}$ be bounded. Then, $f(x)$ is Riemann integrable if and only if the set of discontinuities of $f(x)$ has measure 0.

Thomae's function, being discontinuous only on the rational numbers in $[a, b]$ is discontinuous on a set of measure 0, and is therefore Riemann integrable, according to Lebesgue's criterion. Dirichlet's function, on the other hand, is discontinuous on all $[a, b]$ and is therefore not Riemann integrable according to Lebesgue's criterion, a fact we already know is true.

This brief discussion is meant only as a very broad overview and introduction to an important and beautiful chapter in mathematical analysis that is the natural sequel to this text: measure theory and the Lebesgue measure and integral.

Once we have Lebesgue's measure and integral, we can define the space of Lebesgue square integrable functions over an interval $[a, b]$, $L^2([a, b])$, which is the set of functions such that

$$\langle f, f \rangle = \int_{[a,b]} f^2(x)dx < \infty.$$

Thanks to the Lebesgue integral, this vector space of functions is complete under the mean-square norm $\|f\| = \sqrt{\langle f, f \rangle}$, which means that every Cauchy sequence in $L^2([a, b])$ converges in $L^2([a, b])$, with respect to the metric induced by the norm.

This complete space or Banach space, endowed with the inner product

$$\langle f, g \rangle = \int_{[a,b]} f(x)g(x)dx,$$

which induces the norm $\sqrt{\langle f, f \rangle} = \sqrt{\int_{[a,b]} f^2(x)dx}$ is an instance of a Hilbert space, an inner product space complete with respect to the metric induced by its inner product. The Fourier basis of sines and cosines, adjusted for the interval $[a, b]$, is a countably infinite basis for this infinite dimensional space.

Banach space and Hilbert space are terms from functional analysis, another important and advanced subject in mathematical analysis.

Further Readings

1. *The Elements of Real Analysis*, Robert G. Bartle, John Wiley & Sons, 2nd Edition, 1976
2. *Principles of Mathematical Analysis*, Walter Rudin, McGraw Hill, 3rd Edition, 1976
3. *An Introduction to Lebesgue Integration and Fourier Series*, Howard J. Wilcox and David L. Myers, Dover Publications, Dover Edition, 1995
4. *Fourier Analysis*, T. W. Körner, Cambridge University Press, Reprint Edition, 1989
5. *Elementary Analysis: The Theory of Calculus (Undergraduate Texts in Mathematics)*, Kenneth A. Ross, Springer, 2nd Edition, 2013
6. *Introduction to Real Analysis*, Robert G. Bartle, Donald R. Sherbert, Wiley, 4th Edition, 2011
7. *Real Analysis, Measure Theory, Integration, and Hilbert Spaces*, Elias M. Stein & Rami Shakarchi, PRINCETON UNIVERSITY PRESS, 2005

© The Editor(s) (if applicable) and The Author(s), under exclusive license
to Springer Nature Switzerland AG 2025
A. Portnoy, *Calculus to Analysis*, Synthesis Lectures on Mathematics & Statistics,
https://doi.org/10.1007/978-3-031-69662-6

Index

A
Antiderivative, 55–58
Archimedean property, 4, 5
Archimedes, 96, 101
Area under the graph, 51

B
Banach space, 113
Bessel's inequality, 85, 35
Binary represented, 12
Bisection algorithm, 32, 33
Bolzano, 16–18, 31–33, 104
Bounded or dominated convergence theorem, 93

C
Cantor, 6, 8, 11, 12, 47, 48, 111
Cardinality, 12
Cauchy, 16–19, 38, 51, 53, 68, 92, 93, 112
Cavalieri, 98, 99, 101
Cesàro, 21, 22, 88, 89
Compact, vii, 10, 11, 16, 18, 19, 31, 33, 92, 103, 104
Continuous, 28, 30, 31, 33, 34, 39, 40, 44–48, 53–55, 57, 61, 64, 68–71, 75, 80, 87, 91, 93, 94, 104, 111, 112
Contraction, 37
Contradiction, 6
Converges absolutely, 22, 79
Converges conditionally, 22, 23
Converges pointwise, 59, 68, 94

Converges uniformly, 60, 69, 112

D
Decimal representation, 7
Derivative, 39–44, 46, 48, 54, 55, 57, 58, 69, 71, 92, 102
Devil's staircase, 47
Dido's problem, 95
Differentiable, 39, 40, 43–47, 61–63, 68–72, 75, 80, 92
Dirichlet, 28–30, 54, 66–68, 73, 88, 89, 91–93, 110, 112
Dirichlet kernel, 88, 89

E
Eigenvalue problem, 76
Eudoxus, 96
Evaluation theorem, 55, 56

F
Fejer kernel, 88–90
Fixed point, 34–38
Fourier, 73, 75–81, 83, 84, 86–91, 93, 94
Fourier basis, 77, 78, 84, 86, 87
Fourier series, 73, 75, 78–81, 83, 84, 86–88, 90, 91, 94
Functional analysis, 113
Fundamental theorem of calculus, 55, 56, 58, 61

G
Gabriel's horn, 58
Geometric series, 20, 38
Gibbs phenomenon, 78

H
Harmonic series, 21–23
Heat equation, 75, 76, 94
Heaviside, 28, 29, 109
Heine-Borel, 10, 103, 104
Hilbert space, 113

I
Indefinite integral, 56
Infimum, 3, 4, 33
Inner product, 83, 84
Instantaneous rate of change, 42
Integers, 1
Integration by parts, 57
Integration by substitution, 57
Intermediate value theorem, 31, 34, 103, 104

L
Least-squares approximation, 85, 90
Lebesgue, vii, 86, 93, 94, 107–112
Lebesgue integral, vii, 93, 94, 107–112
Leibniz, 58, 100, 101
Lipschitz, 31, 37, 38, 92

M
Maclaurin polynomial, 62
Mathematical induction, 1
Mean value theorem for derivatives, 47
Measurable sets, 109
Measurable space, 109
Measure, 12, 93, 94, 107–109, 112
Measure space, 109
Method of exhaustion, 96
Metric, 83, 86, 90, 92, 93

N
Natural numbers, 1, 6, 7, 15
Nested, 11, 12, 19, 32, 103, 104
Net distance traveled, 54

Newton, 41, 100–102
Newton's method, 48
Norm, 83

O
One-sided limits, 25, 39, 83, 88, 91
Open cover, 103, 104
Ordered field, 3
Orthogonality, 77, 78, 83, 84
Orthogonal set, 84

P
Parseval's identity, 86, 87
Partial sums, 20, 21, 69, 71, 72, 78–80, 89
Partition, 51, 68, 91, 107, 110
Piecewise continuous, 83
Pointwise limit, 59, 66–68, 71, 92, 93
Power series, 62, 64, 72
Projections, 84
Pythagoras, 2

Q
Quadrature of the parabola, 96, 97

R
Rational numbers, 1, 2, 4–6, 8, 9, 28, 30, 66, 109, 112
Real numbers, 1, 3–7, 15, 19, 23, 25, 28, 53, 65
Rearrangement, 23
Refinement, 51
Riemann integrable, 51, 54, 67, 68, 92–94, 112
Riemann integral, vii, 51, 53–55, 58, 68, 93, 107, 108, 110, 112
Riemann-Lebesgue lemma, 86
Riemann sum, 51, 54
Rolle's Theorem, 46

S
Separation of variables, 76
Sequences of functions, vii, 59
Slope of the tangent line, 42, 43, 55, 57
Supremum, 2, 4, 33, 92

T
Taylor polynomial, 62
Taylor's inequality, 62, 63
Thomae, 28, 30, 54, 112
Trigonometric series, 73, 75, 77, 83, 88, 89, 94

U
Uniform limit, 60, 68, 69

Uniformly continuous, 31, 33, 48, 53, 70

W
Weierstrass, 16, 18, 31–33, 64, 66, 71, 72, 75,
        80, 88, 89, 104
Weierstrass function, 70–72
Weierstrass theorem, 16, 18, 31, 33, 64, 104
Well-ordering, 1